How To Grow Your Own Vegetables

Michael Kressy

Illustrations by Ed Kressy

Introduction by C. L. Thomson
Professor of Vegetable Crops
College of Agriculture
University of Massachusetts

CHL CREATIVE HOME LIBRARY
In Association with Better Homes and Gardens
Meredith Corporation

CREATIVE HOME LIBRARY

Printed in the United States of America

Library of Congress Cataloging in Publication Data

Kressy, Michael, 1936–
How to grow your own vegetables.

1. Vegetable gardening. 2. Berries.
3. Cookery (Vegetables) 4. Cookery (Berries)
I. Title.
SB321.K74 635 72-11963
ISBN 0-696-24600-7

Contents

About the Author

Michael Kressy's gardening experience ranges from growing things in New York City window sills to working on a tobacco and truck farm in New England's Pioneer Valley. Formerly a journalist, Kressy now teaches creative writing at a community college in Massachusetts. A confirmed dirt farmer, Kressy grows much of the food that he, his wife, and two children need, including the crops discussed in this book.

Author's Preface

A large part of gardening is scientific. But the impulse to grow things and persevere must come from the heart and not the head. Most of this book is bare bones information but the ordering of its parts and the manner of presentation are based on two convictions:

that Nature is not the eternal adversary she is often made out to be but the benevolent overseer of life as well as a source of joy for those who are willing to watch and listen. Therefore, the book is predicated on an abiding affection for the outdoors and accedes to Nature the final word on making things grow.

that in spite of the morass of do's and don'ts that often characterize gardening manuals, success in growing things is ultimately a matter of individual creativity. Therefore, the book is intended to encourage that creativity and not in any way to invite conformity of thought or method.

If, after reading these pages, the reader is inspired to set seeds to soil and discover through gardening a vague but comforting link to the universe, the book will have been worthwhile.

How To Grow Your Own Vegetables could hardly have materialized without the patient help of my wife, Jean, who not only kept the din of two growing children at an acceptable level but also contributed substantially to the "After the Harvest" section, including inventing and testing the marvelous recipes in chapter 10. At the same time, any book on gardening must be heavily indebted to the United States Department of Agriculture and its network of state experiment stations. Grateful thanks is also extended to the Fitchburg, Massachusetts Public Library for tracking down source material and to Marion Rahaim for help in preparing the manuscript.

Ashburnham, Massachusetts
October, 1972

Michael Kressy

Introduction

All over America there is increasing interest in growing vegetables, fruits, and herbs in the home garden. Some who are already profiting by this healthful activity were drawn to it by practical considerations—a need for budget retrenchment due to soaring food prices or an awareness of the lack of quality and taste in store-bought vegetables and fruits. Still others wanted to help conservation and the ecology by using "organic" methods of gardening and thus not be heedlessly mining the soil for more than it is worth. Growing fresh food in your own home garden offers a splendid opportunity to meet all of these goals.

Whether you are six or 60, you will be enormously rewarded if you participate in this leisure-time occupation. A well-planned, carefully tended garden provides the table with a copious supply of delicious vegetables and fruits which are rich in vitamins and minerals as well. And gardening means getting away from tension and stress by getting plenty of pleasant exercise in the sunshine and fresh air.

With modern tillage equipment gardening is a fairly easy job that is not too time-consuming. And you don't need a large area, either; a few feet of soil in the sun will be prodigiously productive. Even a window box will grow lettuce, radishes, and other quick crops. Herbs can be planted in pots.

One of the most impressive qualities of *How To Grow Your Own Vegetables* is the concern for the soil and the general environment displayed by the author, Mike Kressy. He suggests many natural and non-hazardous methods of controlling plant pests and diseases. He urges the gardener to become well acquainted with the friendly, helpful insects as well as with the harmful insects so that the former are never accidentally destroyed. Although Kressy stresses biological control of pests and diseases, he realizes that chemical insecticides sometimes have a place and does not hesitate to recommend their use. However, careful application of chemical insecticides is advocated strenuously so that other life is not endangered. In short, good sense is the hallmark of this book.

The food-producing garden is an ideal place for the entire family to get in tune with nature, to find spiritual as well as physical and mental fulfillment. With this book as a guide, the family can experience the miracle attendant upon each stalk pushing through the ground. It is my pleasure to introduce *How To Grow Your Own Vegetables*—a readable, practical, and comprehensive handbook for any and every home gardener.

C. L. Thomson
PROFESSOR OF VEGETABLE CROPS
COLLEGE OF AGRICULTURE
UNIVERSITY OF MASSACHUSETTS

Amherst, Massachusetts

I.

Planting Time – The Groundwork

1.

Plan Before You Leap

Since growing conditions can often spell the difference between large succulent fruit and shrunken or disease-ridden produce, choosing the right spot will help you to get off to a good start. When picking a suitable location for the garden, keep the following factors in mind as you check your backyard:

1. Keep the garden nearby. Ideally the family garden should be no more than a carrot's throw away from the kitchen door. If the family cook can see rich, red tomatoes from the kitchen window, she's more likely to see that they find their way to the dinner table. Close proximity to the house will also help you discourage small animals from nibbling at lettuce leaves. At the same time, weeding and general maintenance will be less of a chore because tools and water hose will be close at hand.

2. Sun, sun, and more sun. The simple truth of gardening is that all vegetables and fruits need ample amounts of sunlight. Generally speaking, the garden should receive unobstructed sunlight all day.

Rich soil, plenty of sunshine, and ample moisture are essential ingredients for successful vegetable and fruit growing.

3. Search out the best soil. Look for a spot where the topsoil—the fairly dark-colored surface layer that is rich in organic material—is deep, friable, and well drained. In some areas the topsoil may consist of only an inch or two of good soil. In others, as on the prairies, it may be several feet thick. Test it. Squeeze a handful of moderately moist topsoil. If it crumbles or at least does not stay in a solid clod when you open your hand, it will probably support crops. Even if it stays compact, don't scrap your garden plans. All soil can be improved and good soil can be made even better. Additions of humus, compost, manure, and other fertilizers worked into the top eight inches of soil will bring it close to ideal.

Even subsoils—those pale-colored layers beneath the organic-rich topsoil—that are tightly compacted and unsuited to growing crops can be brought up to par by additives. Subsoils usually contain valuable mineral elements that plants need, and they are weed-free, too. Subsoils may have a high clay content, which makes them sticky when wet and hard when dry. Additions of peat moss or other humus, as for topsoil, will open up the texture. If there is a great deal of clay, adding sand or cinders will further open it up to give drainage. A winter cover crop ("green manure") that is turned under each spring for the first few years also will help to improve the soil (see page 24), and within a few years you will have superb soil for growing vegetables.

Hardpan is a problem in a few areas. This is a layer of dense cementlike soil that prevents drainage and won't permit root penetration. Plants suffer and become sickly. If the hardpan is two feet or more below the surface, few ill effects will be felt, provided there is some drainage. If it is near the surface, other measures may be needed. Breaking up hardpan to get to the subsoil below is hard and expensive. It may be a better plan to build up beds on the surface. This can be done by edging beds with railroad ties, staked boards, or masonry to hold topsoil and compost to a depth of 8 to 12 inches. The patterns of such beds make interesting visual effects, too.

4. Give trees a wide berth. The expansive root systems of trees can effectively drain the surrounding soil of available nutrients. As a general rule the garden should not intrude inside the drip line of a tree (the outer limits of the leaves).

A tree's root system extends to the drip line—an imaginary line drawn where rain dripping off the outermost leaves strikes the ground. Plant crops outside drip line so that nutrients in soil will go to the vegetables instead of the tree.

5. Look for a southern exposure. Because ground that slopes slightly to the south (but with no more than a 1.5 percent grade) gets more direct rays from the sun for longer periods, plants have a much greater chance of surviving the ravages of frost in areas where cold nights are commonplace.

6. Choose a high area over a low area. In spite of an apparently still atmosphere, the air temperature constantly shifts as the day progresses, even on a June day. Cold air tends to move into lower areas. As a result, in frosty cold zones vegetables or fruits planted in depressions will be forced to struggle with frost more often than will their high-ground counterparts. A high flat spot or a slope rather than a flat low site will always be the better choice.

WHAT TO DO IN CRAMPED QUARTERS

Naturally the peculiarities of your own backyard will make it easy for you to meet some of these requirements and perhaps impossible to meet others. Don't despair, however, if the smallness of the yard, the placement of trees and shrubs, or the location of the dwelling seems to have left no space available, making a vegetable and fruit garden seem nothing more than a dream. There are a variety of ways to rescue nooks and crannies that have succumbed to blown-in refuse and spiders, or neglected retaining walls that have all but disappeared beneath a carpet of weeds. Keep in mind three approaches to squeezing luscious crops into limited spaces: (a) incorporate them into the landscaping scheme; (b) provide special areas for specific crops; (c) use space-saving gardening techniques. For example:

Fitting Crops into the Landscape All too often the esthetic qualities of fruit trees and vegetable plants are overlooked in favor of their practical value. Many varieties not only provide edible fruits but also possess unusually shaped or colored leaves. Others grow on gracefully cascading vines that can add color and beauty to any wall or fence. Give these ideas some consideration before saying no to a particular vegetable or fruit:

FLOWER BEDS AND BORDERS—Swiss chard, lettuce, chives, parsley, thyme, and rosemary all make exquisitely delicate edging plants for walks or flower beds. Large crops such as cabbage and eggplant add interesting leaf textures to the flower display. Eggplant possesses especially eye-catching velvety leaves in addition to delicately tinted lavender blossoms. Other crops can add a much-needed color note elsewhere. Rhubarb, for example, has valentine-red stalks; red cabbage offers a deep purple tinged with green. For intriguing leaf textures try the loose-leaf varieties of lettuce or try asparagus, which looks like feathery wisps of green cotton when full grown. Corn, because of its size, can give a prehistoric flavor to the landscape as well as provide a pleasing backdrop for lower-growing flowers. Strawberries, on the other hand, make a perfect ground cover.

CONTAINERS FOR PATIOS, TERRACES, OR DECKS—Just because the yard is filled with necessary objects or is nonexistent is no reason to dismiss vegetable and fruit growing out of hand. Raised beds constructed of wood used in imaginative ways can provide the family with garden treats through the entire summer. Cherry tomatoes, a must for salads, will grow lavishly in hanging baskets lined with moss

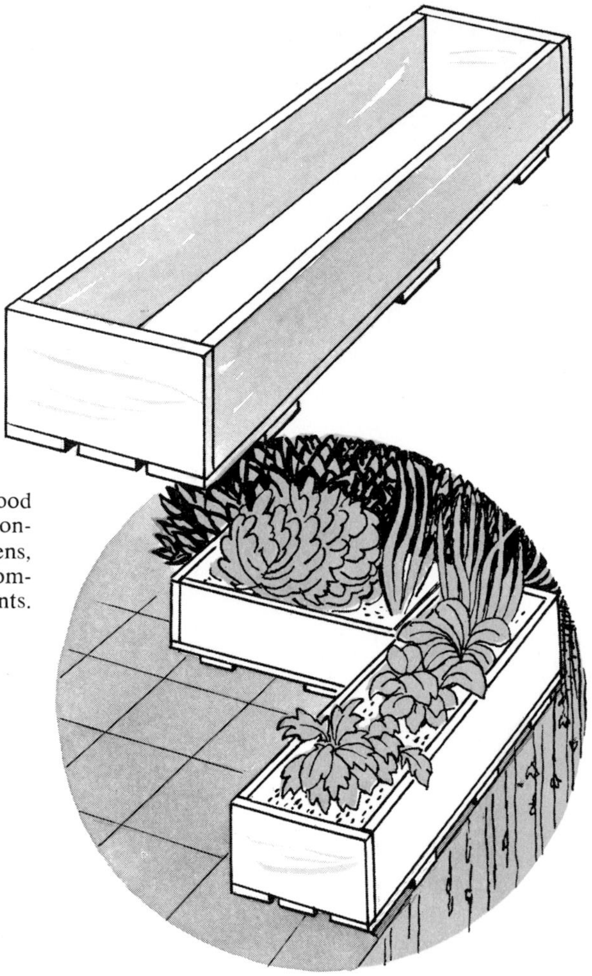

Wooden boxes of preservative-treated wood at least 6 to 8 inches deep make excellent containers for plantings of herbs, salad greens, or strawberries. Several boxes can be combined and placed in attractive arrangements.

and filled with soil. The cool green cascade of cucumber leaves flowing out of a barrel filled with good loam mixed with well-rotted or processed packaged manure can be a spectacular addition to any terrace or patio. Brick or stone terraces with small plantings here and there to break up the monotony have advantages all their own. The stone serves as a natural mulch, keeping in moisture and preventing weeds from soaking up valuable soil nutrients. Raised beds need extra dousings of water but provide excellent drainage conditions for plants. Strawberries do especially well in beds lined with stone or with cedar or redwood.

WALLS, FENCES, AND RAILINGS—A sun-drenched wall or fence can provide excellent growing conditions for warmth-loving crops such as melons, cucumbers, and squashes. Cucumbers are especially cooperative because they can be trained to grow up against a wall or fence, producing clean unblemished fruits. If part of your lawn consists of a stone outcropping, turn it into a pleasingly aromatic herb garden, or plant trailing-vine crops such as melons or pumpkins. The stone will help preserve moisture and at the same time keep the produce clean and free of soil-borne diseases. Beans or peas will grow happily against railings or porch pillars. Vegetables can be as effective as costly nursery shrubs in breaking up the harsh lines of a building. Then, of course, there is also the bonus of nutrient-rich fresh produce.

Keep in mind the space-yield ratio. Some vegetables require much more room than others per pound of produce. For example, during a season sweet corn will produce only 15 pounds per 100 feet of row while carrots or tomatoes will top 100 pounds in the same amount of space. The trick is to become as familiar as you can with the growing habits of the various vegetables and fruits by studying their characteristics listed in chapters 6 and 7. Then pick those that offer the greatest rewards in the least amount of garden space. In general, when space is limited consider these gardening tips:

> Swiss chard is a better choice than spinach because it offers continuous picking throughout the growing season. For the same reason, the loose-leaf varieties of lettuce make more sense than the heading types.
>
> Go vertical wherever possible. Cucumbers, peas, and pole beans are natural climbers and will not object to growing on trellises or along fences.
>
> Tomatoes, eggplant, and pepper are bush plants that require very little room but produce abundant yields.
>
> Avoid the large plants like corn or potatoes. If supports are not available, then vine crops such as squash and pumpkin should also be passed up in favor of leaf or root crops.
>
> The vegetables in the onion family (chives, garlic, leeks, onions, shallots) require little space, are hardy, and will grow in a wide variety of soils.

COOL SEASON AND WARM SEASON CROPS

Vegetables vary according to their ability to withstand cool temperatures. A hard frost may actually improve the flavor of Brussels sprouts but wreck a tomato patch. Get to know the vegetables you want to grow—it will make planning that much easier as well as prevent possible failures because the wrong thing is planted at the wrong time.

The first thing to be considered is how adaptable a particular vegetable is to cool temperatures. Cold resistant crops, such as cabbage, peas, or beets, are of special value to the gardener because they can thrive in the spring and fall or throughout the winter in southern regions. There are other important differences too. Cool season crops are generally smaller in overall plant size, have shallower root systems, and germinate at cooler soil temperatures. Warm season vegetables, on the other hand, grow sluggishly and produce low yields if exposed to constant low temperatures. When planning, give tender (cold sensitive) and hardy (cold resistant) vegetables separate locations. Then in spring only the patch reserved for the hardier types need be prepared. Prepare the rest of the garden after danger of frost has passed. Study the following lists carefully before finalizing your garden plan.

Crops like peas and lettuce actually prefer cool weather while others such as beans, eggplant, and corn demand warmth. Gardening chores can be simplified if tender and hardy vegetables are grouped together.

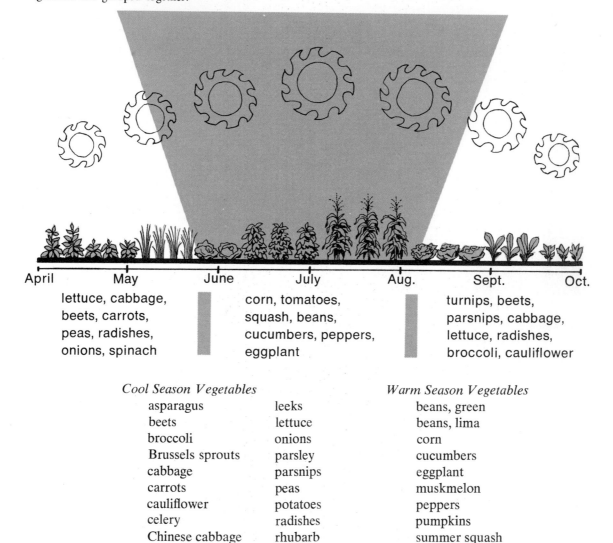

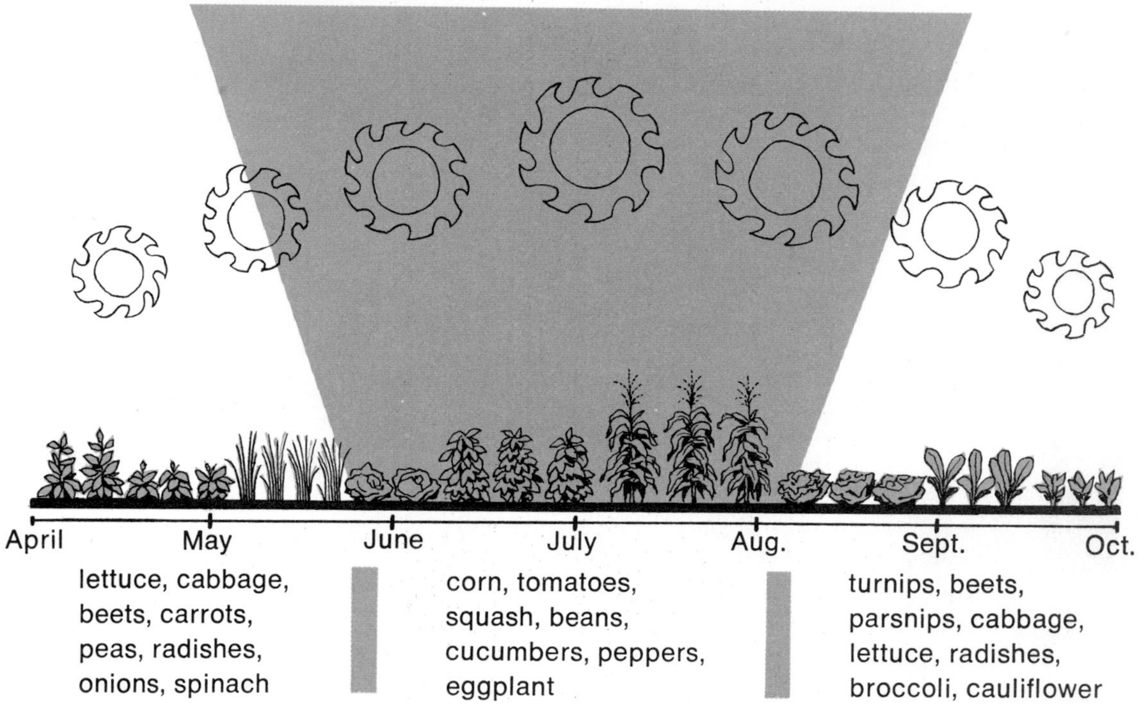

Cool Season Vegetables		*Warm Season Vegetables*
asparagus	leeks	beans, green
beets	lettuce	beans, lima
broccoli	onions	corn
Brussels sprouts	parsley	cucumbers
cabbage	parsnips	eggplant
carrots	peas	muskmelon
cauliflower	potatoes	peppers
celery	radishes	pumpkins
Chinese cabbage	rhubarb	summer squash
Swiss chard	spinach	winter squash
chives	turnips	tomatoes
		watermelon

PREPARING A GARDEN PLAN

The planting demands of seeds and the onrush of spring weather can induce a panic that threatens the success of the entire project unless a well-thought-out garden plan has been prepared beforehand. A pencil, graph paper, a reliable seed catalog, and a list of vegetables and fruits are all that are necessary to work out your plan. First check the growing characteristics of each vegetable and fruit (see chapters 6 and 7). Is the plant perennial or annual? How long a growing season does it require? Can succession plantings after early harvesting be made? Is it hardy, tender, or frost tolerant? What amounts should be grown for the number of people in your family?

Once these questions have been answered, arrange the garden according to these general principles:

1. Put perennial crops (plants that come up every spring without replanting) together at one side where they can flourish unmolested by plowing and cultivating equipment or by the digging each year for annual crops.

2. Keep together hardy root crops that can be planted as soon as the ground permits working. Beets, onions, carrots, parsnips, and all other vegetables grown primarily for their roots need a well-prepared soil free of stones and pieces of sod.

3. Arrange according to height. Tall plants like corn and pole beans will cast meddlesome shadows over ground-hugging lettuce or beets if not kept to one side of the garden.

4. To conserve labor, keep rows as straight and regular as space permits. The vegetables, of course, can't tell a straight row from a crooked one, but the gardener's tasks of cultivating and other garden chores are made much easier.

5. If you possess enough yard space to juggle the location of the garden, try to position the rows so that they run north and south. This way all the plants will get equal doses of life-giving sunlight. But if the garden is on a slope running from north to south, the rows will have to run east and west to prevent erosion.

6. Get the most out of the space available by mixing crops together, either in the same row or in alternate rows. Lettuce, a quick-maturing crop, is an excellent candidate for intercropping with slower-growing cabbage, broccoli, eggplant, or tomatoes. Either a continuous row can be squeezed into the space between two rows of a large crop or individual plants can be placed within the row.

7. Succession-plant—two, three, or more seedings of small amounts of the same vegetable—at regular intervals to insure a constant supply of short-season or quick-maturing vegetables. Plan on a ten-foot row of lettuce, radishes, Swiss chard, or spinach with space nearby reserved for a second and third row to be planted at seven- to ten-day intervals. When row one is depleted, row two will be reaching peak flavor, while row three will be two to three inches high. At this point row one can be raked smooth and replanted.

THREE GARDEN PLANS

The garden plans shown here will give you some idea of how your vegetables and fruits can be arranged. Naturally, your own plan will be varied to fit the twists and turns of your backyard, as well as your food likes and dislikes. The size of your garden will depend on the time you have for gardening and the space. The New Homestead plan, for example, calls for a plot 25 by 40 feet and will probably require a little more care than weekend gardening will allow. Remember, it's much better to start small and expand later than to have your gardening enthusiasm dampened at the outset because you attempted a too big and complex garden.

Mini-Garden

No matter how small your home lot may be, some space can usually be found for a vegetable garden. Only 6 feet by 8 feet, the Mini-Garden introduces you to the gourmet world of fresh vegetables without demanding too much time—an hour or two a week is plenty. If mulches, either natural or plastic, are used, the Mini-Garden will be practically maintenance free.

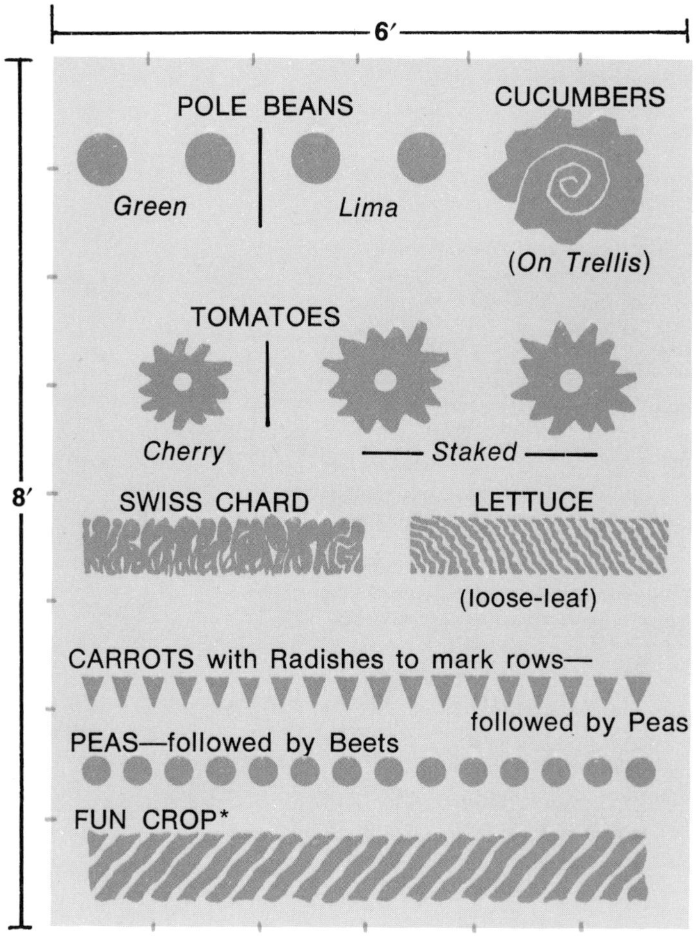

*Chinese cabbage
celery
leeks
muskmelon (on trellis)
watermelon (on trellis)
etc.

Beginner's Special

The Beginner's Special is for those who have dreamed about a vegetable garden for years but who have never lifted a spade and may be reluctant to dig up more than they can hoe. If you plan carefully, allowing for succession planting of quick-maturing vegetables and intercropping of slower-growing ones, you'll have enough fresh produce for freezing and canning as well as for the dinner table.

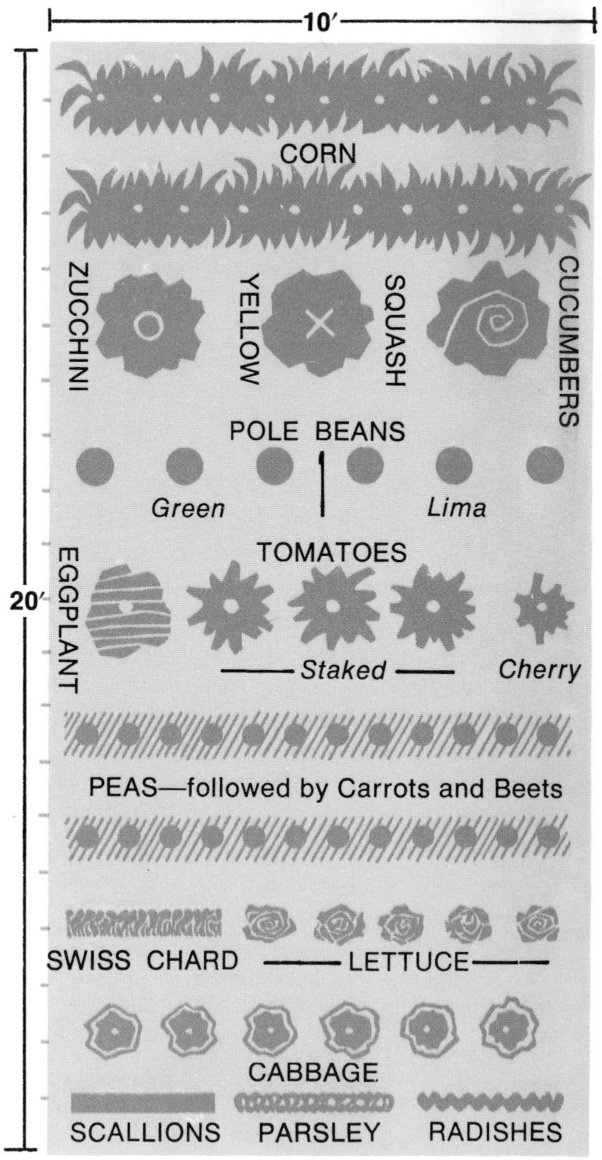

New Homestead

This plan enables an average family to grow a lot of the produce it eats. Much larger than the preceding plans, the 25 feet by 40 feet New Homestead requires more working time as well as more space. But the rewards of fresh asparagus and rhubarb make additional maintenance chores more than worthwhile. If you decide on a garden this size or larger, you might consider purchasing a rotary tiller.

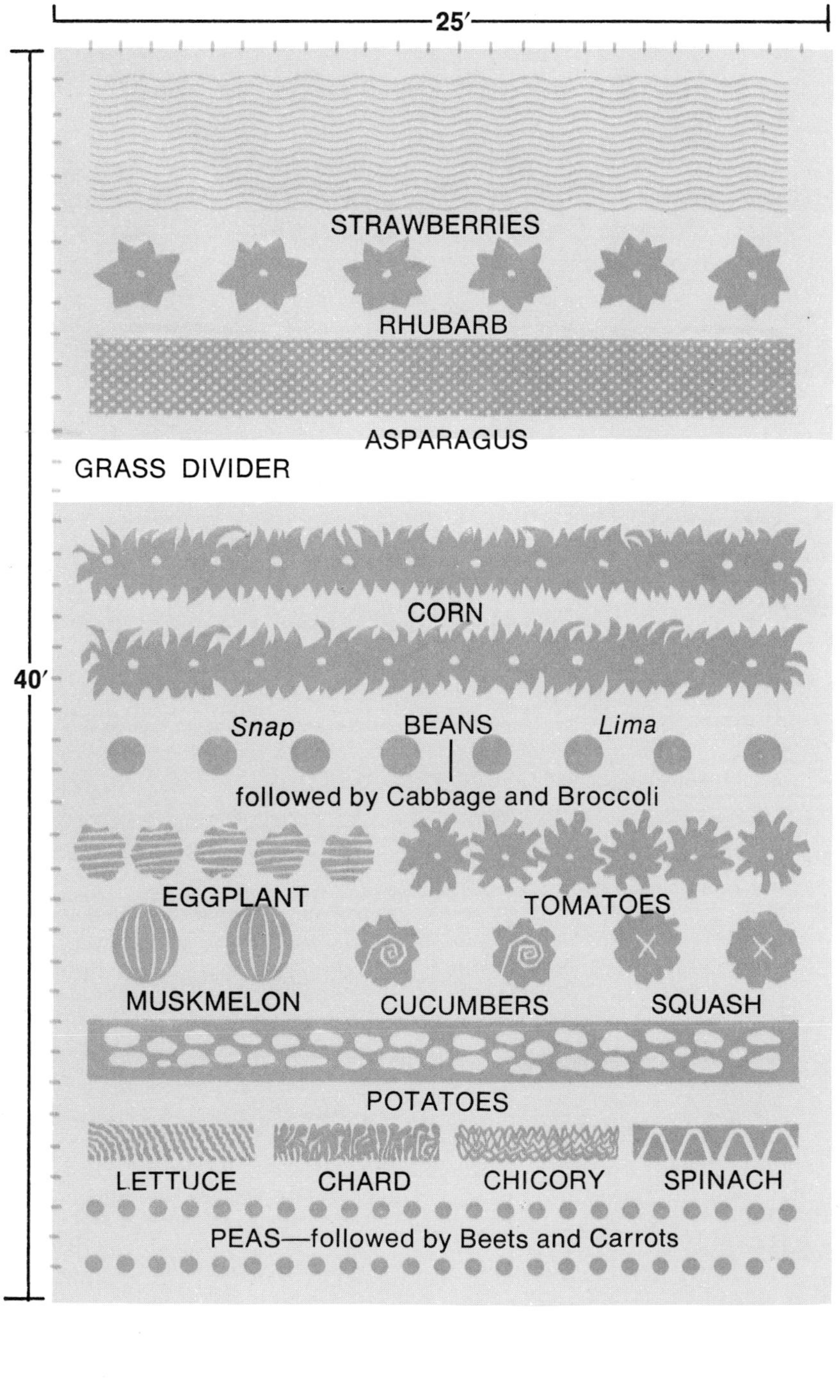

KEEP NOTES ON WHAT YOU DO

A large part of the excitement and creativity of gardening comes in fitting garden practices to your particular soil and climate conditions. Friendly kibitzing from friends and neighbors may be helpful at times, but nothing matches the lessons you learn from experience. But experience soon fades into memory, and memory for most of us fades into forgetfulness if things aren't written down. Develop the habit of summarizing in a diary or notebook what you have done and whether or not it has proved successful. After several seasons your notes can be translated into a personal garden calendar that will allow you to operate like an old pro. A notebook entry might look like this:

CROP	VARIETY	AMOUNT PLANTED	DATE PLANTED	DATE AND AMOUNT HARVESTED	COMMENTS
Lettuce	Salad bowl	2/10ft rows	April 15	May 28- first day of harvest	Succession crop, interplanted with cabbage and Brussel sprouts

The advantages of note taking far outweigh the small amount of time necessary to write down what you do and how you do it.

Mistakes are nipped in the bud, before they have a chance to proliferate and cause grief a second time around.

Actual dates of plantings and harvestings give the home gardener a chance to plan his time wisely, to avoid the experience of the family being inspired by a beautiful spring day (probably premature) and dashing forth to garden without any clear idea about what to do.

Keeping track of varieties makes it possible to experiment with other, newer varieties and compare results.

General practices gleaned from your notes are custom made for the special conditions that prevail in your backyard.

Keep each year's plan with your notes. (See pp. 12 and 13 for plan diagrams.)

THE BASIC TOOLS OF THE GARDENING TRADE

The greatest surprise in store for the budding gardener is how few tools are necessary to get a garden under way.

Spading Fork Made of heavy steel, usually with four tines, the spading fork is essential for turning, breaking up, and preparing the soil for crops.

Hoe The simplest and perhaps the most basic tool of all, a hoe is a gardener's constant companion because of its usefulness in cultivating and breaking clods, loosening the soil, and removing weeds.

Steel Bow Rake Resembling an oversized comb on a long handle, the steel rake is indispensable for preparing a level seedbed free of rocks, debris, sod lumps, and broken roots.

Twine Perhaps the most unglamorous tool of all, twine in many ways contributes the most by preventing the garden from turning into a jungle of vines and plants. Straight, easily maintained rows are a certainty when twine is stretched between stakes and pulled taut, and the furrow made with the line as a guide.

Watering Devices Since the weather doesn't always come up with all the moisture healthy crops need, the gardener must be ready to lend a hand. A watering can, a garden hose, or a sprinkling device will help to carry vegetables and fruits through an unexpected dry spell. In some dry climates, of course, daily watering is a necessity.

Hand Sprayers Another tool high on the priority list is a hand sprayer or duster for combating harmful insects and diseases. Naturally, as the gardener's enthusiasm grows, more equipment may be needed.

Other tools, while unquestionably valuable in terms of time and effort saved, are essentially frills. A garden tractor or power cultivator is almost a must, however, for larger gardens.

Whatever accessories you purchase, keep a sharp eye out for quality. The investment is so small that it's worth a few extra cents to own well-constructed tools, which are not only longer lasting but also safer to use.

Tools with oak handles will cost more but are cheaper in the long run because they hold up better, resisting bending, breaking, and splintering.

Look for solid heavy-gauge steel on metal parts. Heavy steel stays sharp and doesn't bend. Rivets holding steel to wood should be strong and solid. A good tool will probably be slightly heavier, but a few extra ounces are a small concession to make when it comes to dependability and ease of use.

When purchasing power equipment be sure the machine matches the job you want it to do. A small riding lawn mower just isn't designed to till the soil or haul heavy loads, nor is it practical for very small gardens.

Jump at the chance to pick up secondhand tools or hand-me-downs from a gardening relative. Often old tools not only are well made but also have a pleasant nostalgia about them which makes them a joy to use.

Try to resist the temptation of fancy gadgets. Advertisers will always attempt to convince you that the most recently developed gimmick is indispensable to successful gardening, but this is not always true. The simplest tools are easier to maintain and usually last longer.

When considering power equipment, keep safety foremost in your mind. Moving blades, cutters, or forks should be housed in protective coverings well away from the operator's hands and feet. The ideal machine has a separate clutch arrangement for the blade parts so that cutting or tilling can be stopped immediately without necessarily shutting off the engine.

TAKING CARE OF TOOLS

When winter has finally ebbed and the ground is begging for seeds is no time for tools and equipment to break down. Be ready for the delightful winds of spring by getting your tools in shape beforehand. Winter is the ideal time for these chores. Sharp, well-maintained tools also save time and prevent injuries. A dull hoe, for example, requires more hours of hacking at the earth, not only taking its toll of your energy but also presenting a real danger to the root systems of crops. A few hours in the shop or cellar during the winter months will put your equipment in ideal condition for the jobs ahead.

ENGINES (internal combustion) should be drained of oil and new oil put in. This old oil strained through cotton cloth is perfect for coating the handles and steel parts of tools (see below). Gasoline should not be allowed to stand in tanks over the winter, since sediment can form that will foul the carburetor. Simply run the engine until the gas supply is exhausted. Remove the sparkplug, and pour in a tablespoonful of oil to coat the cylinder wall and piston. Before replacing the sparkplug, give the engine a few turns by hand to distribute the oil.

HAND TOOLS will last indefinitely if kept free of dirt and rust. In the fall, clean tools thoroughly with water and a steel-wire brush. Soil sticking to tools will cause them to rust, which in turn will make them more difficult to use. Make whatever repairs are necessary. If the handle appears to be splitting, wrap wire tightly around the damaged portion to keep the tool together for another season. Handles with splinters can obviously make gardening a painful experience. Sand the wood with a medium-coarse sandpaper until smooth. As a final step, coat the handle and steel parts with a light covering of reclaimed oil from the garden tractor or lawn mower, removing excess oil with a clean cloth.

METAL WHEELBARROWS, CARTS, AND FERTILIZER SPREADERS should be rinsed thoroughly after each use and wiped with a cloth dampened with oil. Oil and grease the wheel axles and other moving parts before storing. Metal that has been exposed by chipping or denting is a perfect place for rust to gain a foothold. These damaged places should be sanded with emery cloth and given a coat of enamel as soon as possible.

2.

The Soil - Where It All Begins

The sun is either in or out, and except for putting the garden in a good spot there is little that can be done about it. By the same token, if you have been cheated out of adequate rainfall the only recourse is to use a watering can or garden hose. But soil is different. With a thimbleful of understanding of what soil is and how it works, a sand pit can be turned into something like the Hanging Gardens of Babylon. All it takes is enough grasp of the basics to do the right thing at the right time in the right amount.

A QUICK LOOK AT SOIL

At first glance the layer of topsoil that is spread like a thick film over the earth may be dry and lifeless. But a good topsoil supports life—all life, including our own. For plants it will supply four crucial things:

1. A fertile medium for root growth and expansion, which not only holds plants upright but also allows for a free flow of air to reach the roots.

2. A spongelike material that allows excess water to pass through it freely and at the same time retains just enough moisture to keep plant roots content, enabling them to take up nutrients in solution.

3. A storehouse where nutrients vital to healthy productive plants can remain continually available to roots.

4. A natural laboratory where soil organisms can carry on the important business of breaking organic matter down into a form usable by plants.

THE PHYSICAL MAKE-UP OF THE SOIL

If you were to take apart a handful of soil, the chances are you would discover a remarkable collection of pebbles, sand, something resembling flour (silt), lumps or clods, old plant roots, a somewhat fibrous dark material (humus), dead leaves and twigs, and, finally, a few ants, earthworms, and other creeping things. Believe it or not, all of these put together make up only 50 percent of the volume of soil. The other half consists of empty spaces between the particles, pores through which the water and air so necessary for healthy crops can penetrate. But what materials make up soil? How does a good soil differ from a poor one?

These are the basic elements of good soil:

SAND feels gritty when a sample of soil is rubbed between the fingers. The coarsest of soil components (except for pebbles and stones), sand contributes to good drainage. But, like sand on a beach, it is a poor retainer of moisture. Too much coarse sand in a soil means that organic material will have to be added.

SILT feels something like talcum powder and will smear like finger paint. In terms of moisture and nutrient-holding qualities it is halfway between sand and clay.

CLAY, fine particles individually invisible to the naked eye, can be a problem if present in excess amounts. Soil with too much clay holds puddles and tends to dry into a hard, solid surface, which sooner or later shrinks, displaying fissures and cracks. If plowed when wet it will form into small impenetrable lumps and become almost useless for growing things. It's possible, however, to redeem clay soil for a garden area by incorporating sand, peat moss, compost, or other conditioning materials in the soil.

HUMUS is the most important element of all because it alone gives soil the sponge-like quality of soaking up and storing nutrients and moisture. It is gardener's gold. If your soil appears dark brown or black, chances are you're the proud owner of humus and are destined to produce bushels of delicious fruits and vegetables. Don't panic if humus is scarce. It can easily be supplied by copying nature's breakdown process through composting (see below).

If everything is mixed together in the right proportions and plenty of earthworms and other living organisms are present, the soil will be crumbly, of good structure, and probably a trifle moist. Soil that has everything going for it is said to be "friable"—open textured and easy to crumble—and is termed loam. Needless to say, such a soil will produce excellent crops. All you have to do is dig.

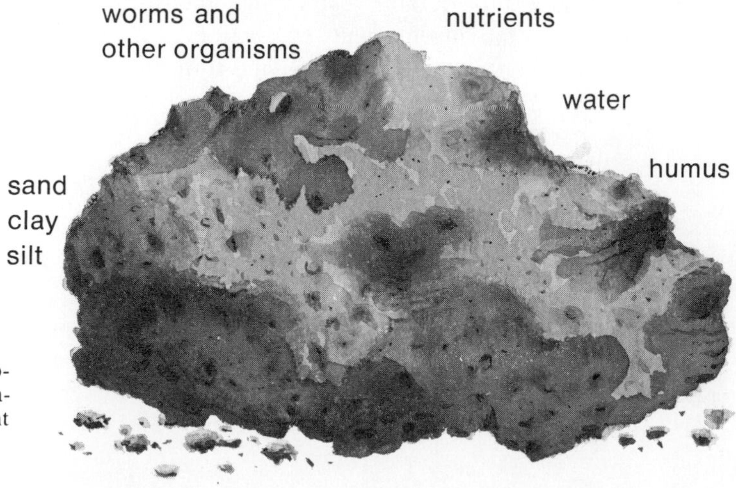

When a soil has good structure it will provide vegetable and fruit crops with a continuous supply of moisture and vital plant nutrients. Such a soil is called loam.

A poor soil, on the other hand, tends to become compact, forcing plant roots to grow near the surface where they are easy victims of drought and hot weather. Drainage will be either too rapid, leaving plants high and dry, or too slow, allowing the pore spaces to fill with water and asphyxiate the oxygen-demanding roots. If in doubt about your soil, keep an eye out for the telltale symptoms: puddling, crusting, shrinking and cracking, soil runoff after a rain, or the dry dusty conditions associated with too-open sandy soils.

If soil is too coarse, water drains off rapidly, depriving plants of adequate moisture.

A crumbly, "friable" soil permits root growth and retains moisture plus nutrients.

A compact soil stunts roots and resists moisture, resulting in spindly crops and poor yields.

HOW TO IMPROVE SOIL STRUCTURE

It should be clear by now that soil is anything but "dirt." The next step is to determine the type of soil you have to work with. Go out and poke around your backyard to see what the soil is like. Let it run through your fingers. Squeeze some on your hand. See whether it is gritty or smooth, moist or dry, light or dark. If it looks as if it needs help, there are a variety of inexpensive, effortless ways to give soil the boost it needs.

COMPOST: WHAT IT IS

The composting process is nothing more than a backyard imitation of something that has been going on in nature long before man stepped into the picture. Next time you're in a field or in the woods, scratch at the soil and you will find a spongy layer of old leaves, partly rotted twigs, grass, and other organic debris covering the soil. If you were to use a magnifying glass you would discover that the deeper it is, the finer the material becomes and the more difficult to identify. This is nature's way of manufacturing nutrient-rich loam. Tiny bacteria with insatiable appetites wait in the soil for digestible organic matter to appear. When fallen leaves, dead branches, and spent plant stalks become compacted on the ground and moisture is added, the bacteria spring into action. In effect, what goes up from the soil to form plants must eventually come down and return to the soil.

The shaggy heap of "refuse" concealed in a shady corner of your neighbor's yard is not a monument to his laziness but is actually a controlled version of what happens naturally in the woods. In constructing a compost pile the gardener

What happens in nature is imitated in the composting process. Falling leaves and other debris form a natural mulch which soil organisms "digest" and incorporate into soil. Plant roots absorb valuable nutrients and transmit them to plants.

artificially creates conditions that lead to the rapid breakdown of biodegradable or organically-occurring matter. The result is open, nutrient-rich humus, ideal for improving soil or for spreading around plants. If all requirements for composting were met and conditions were perfect, you could have compost in three weeks. A more realistic estimate is from three to five months. What materials you use and how you use them depend on where you live, what you have at hand, and how much work you are willing to do. But the important thing is to remember the basic requirements.

Air In order for the microbes to complete the tearing down process they need proper aeration. Turning the pile with a pitchfork or spading fork not only will insure even decomposition but also will allow air to permeate the mixture. Turn the pile twice a season—in early spring and early fall. If turning strikes you as too much work, you can insert vertically into the pile at the start of the season one of the following: two perforated plastic drainage pipes; two columns made from quarter-inch wire mesh; a six-inch square, open end box which is as deep as the pile with holes bored in the sides.

Moisture A dry compost pile will go nowhere. On the other hand, a soggy pile will become inert and might possibly give off offensive odors. A moisture content of between 50 and 70 percent is about right. It's best to sprinkle each layer of refuse with a fine spray from the garden hose as you add it to the pile. Try not to overwater so that water runs out the bottom of the pile, carrying valuable nutrients with it. The material should be moist, not wet.

Nitrogen Adding nitrogen is the small subsistence payment made to the hard-working microbes, which consume prodigious amounts of it in the course of wrecking organic matter. Good sources of nitrogen are cow, chicken, or horse manure, available from dairy or horse farmers or in packaged form at garden supply centers. Dried blood, cotton and soybean meals, or fish scraps are excellent organic sources. If none of these is at hand, a commercial fertilizer (a 5–10–10 formula) may be added in a fine sprinkling over each layer as it is heaped on the pile. (See page 28 for explanation of fertilizer code numbers.)

Heat A compost pile in high gear will reach temperatures of 150 to 175 degrees. The more active the microbes, the warmer the pile. If you don't believe it, see for yourself by poking a crowbar into the middle of a working compost pile; wait a few hours and then see if you can grip the crowbar without flinching. This is also a good way to determine whether the breakdown process is going well. Not only does heat accelerate the decomposition process but it also helps to produce a sterilized soil by killing harmful bacteria, insects, and weed seeds.

Size of Particles Almost anything natural can be fodder for the rampaging microbes, but it must be decomposable and fairly small in size. Kitchen waste, lawn clippings, sawdust, garden refuse, leaves, waste from wineries and breweries, animal wastes—all are good candidates. But coarse materials like cornstalks should be avoided except for a first layer because they are difficult to keep moist and allow too much air to enter the pile and cool it. The finer the material, the quicker will be the decomposition.

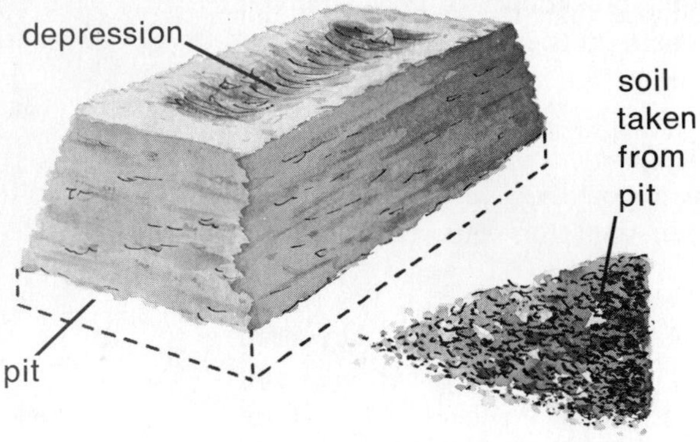

For a nutrient-rich homemade soilbuilder, try establishing a compost pile. Dig a shallow pit and spread in it a 10-inch, even layer of lawn clippings, garden refuse, kitchen waste, and any other biodegradable material. Cover with a thin layer of horse, cow, or chicken manure (or 5 pounds of a 5–10–10 commercial fertilizer) and then with 2 inches or so of the soil taken from the pit. Repeat this layering process as you gradually build up the pile over a period of time. If moisture, air, heat, and organic matter are present in the right amounts, you'll have rich compost in a matter of months. The compost pile will be kept neat if contained with chicken wire and boards.

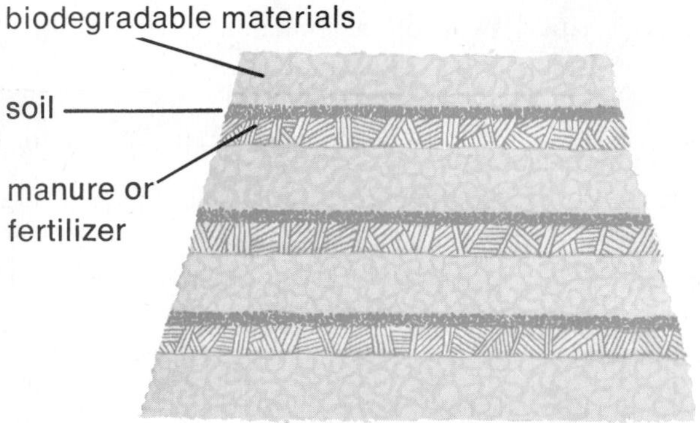

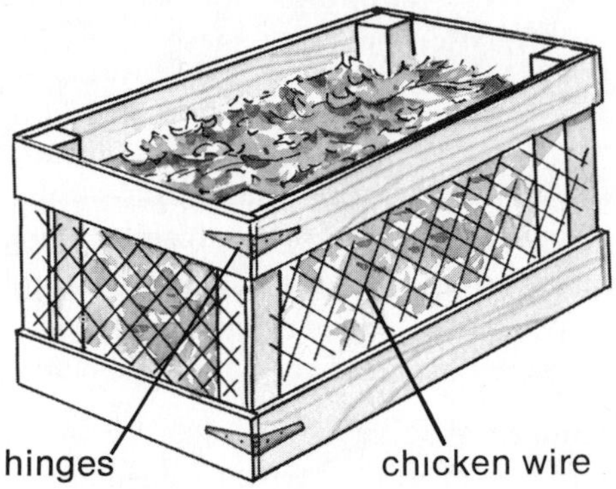

HOW TO GET THE COMPOST STARTED

A compost pile out of sight will probably be out of mind as well and will not get the attention it needs. Find a spot reasonably close to the garden (and lawn) and within garden-hose distance of a water faucet. If possible, locate it in a shady area so moisture will be retained. Once a convenient location has been found the rest is simple.

1. Dig a shallow pit no more than six to eight inches deep and lay the soil aside. Some kind of retaining fence or wire mesh will help to keep the pile within manageable bounds and maintain compaction. Old doors, chicken wire, railroad ties, or cement blocks are all suitable. But remember microbes need air in order to carry on the digesting process. Allow for adequate ventilation by leaving ample space between cement blocks or whatever wood you use. Wire mesh and chicken wire are excellent materials because they keep material compacted and allow for the even distribution of air.

2. Gather together the garbage from the kitchen, lawn clippings, flower bed and garden trimmings. Since the diet of a microbe has almost no restrictions, don't hesitate to throw anything organic into the compost pile. (Stones, tin cans or bones, plastic or paper cartons, metal, and wood should be excluded, of course.) Spread the refuse evenly over the entire pit area to a depth of eight to ten inches. As layers are added, try to shape the pile so that the sides slant

inward. When it is finished (and it shouldn't be more than five feet high at this point) leave a shallow depression in the top layer so that rain will be encouraged to soak into the pile.

3. Moisture is crucial for decomposition. If moisture is not kept at a reasonable level, some of the material may become "fire fanged," or scorched, and useless as a soil amendment. The best time to apply water is after each layer of refuse has been added. Use a fine spray from the garden hose to make sure the pile is moistened thoroughly. In dry climates you can cover the pile with plastic to keep moisture in.

4. Give the heap a shot of nitrogen as you build it. Without nitrogen the microbes will diminish and eventually disappear. A thin layer of either horse, chicken, or cow manure (enough to cover the pile) will make commercial fertilizers unnecessary. For city gardeners fresh manure is not always easy to come by, since farms are becoming fewer and fewer in number. Dried and processed animal manure, however, is available in packaged form from garden supply centers. Other organic wastes such as dried blood, soybean or cottonseed meals, and fish scraps are equally rich in nitrogen and will keep microbes well fed. Barring these naturally occurring materials, a commercial plant food high in nitrogen will serve the same purpose. For an average-size compost heap (say five feet square and four feet high) about 25 pounds of a 5–10–10 mixed fertilizer distributed in five applications of five pounds each will be plenty. A pound or two of lime will keep the pile from becoming too acid.

5. Cap the pile with a two-inch layer of soil. The addition of soil will introduce microbes, keep loose materials from blowing away, and at the same time prevent strong winds from disturbing the top layer of material. Heat and odors will also be kept inside the pile. But remember, only a poorly constructed compost heap will give off disagreeable odors. If conditions are anywhere near perfect, the process will be short and sweet.

6. The pile has to "breathe." The best way to guarantee air penetration is to turn the pile every three or four weeks. Try to work the inside material that is more decomposed to the outside and put the raw material at the edges to the inside. This will hasten decomposition and help to produce a uniform compost.

CANDIDATES FOR COMPOSTING

This list is hardly exhaustive but it will make you a decent scavenger and offers a guide for finding additional materials. Try to avoid animal fat, bones, and other wastes that, although definitely organic, take forever to break down.

Leaves from maple, oak, birch, willow, other trees
Pine needles
Weeds (preferably before they set seeds)
Lawn clippings
Small trimmings from bushes or hedges
Spoiled vegetables or fruits from the garden
Sawdust
Horse, cow, chicken, or sheep manure
Manure from pets
Cornstalks (if well chopped up)
Hay
Garbage from the kitchen
Small wood chips

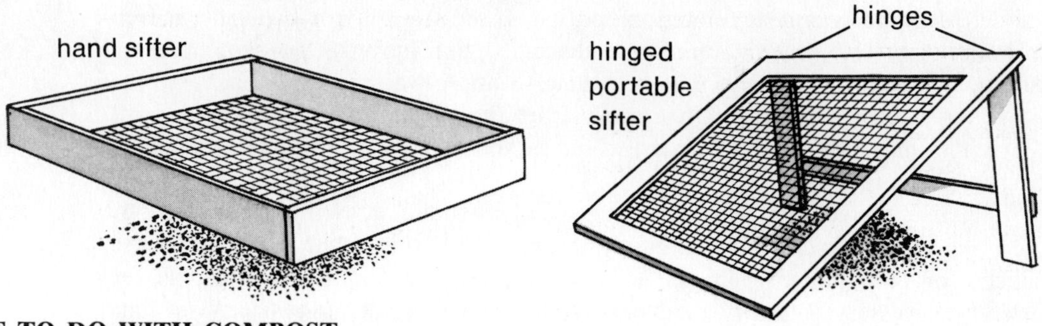

An excellent tool for separating coarse compost from finer materials is a ¼-inch wire mesh fastened to a wooden frame.

WHAT TO DO WITH COMPOST

When the compost has broken down sufficiently it is a rich dark color and crumbles easily in the hand. Now it is ripe for use. The first step is to separate the material into two grades—a fine grade for starting seeds, transplanting, or working into garden soil and a coarse grade for mulching. Wire mesh nailed to a wood frame makes an excellent screen for sifting compost.

As a Mulch Compost that has not yet thoroughly broken down into fine particles is perfect as a mulching material for keeping down weeds and retaining moisture in the soil. For a complete picture of what mulch is and how it works, see chapter 4.

As a Starting Soil Seeds planted indoors will develop into healthy, strong seedlings when started in flats or containers of fine compost. For ways to start seeds, see chapter 3.

As Synthetic Manure If properly managed, the compost pile should produce a soil conditioner that is better balanced than and just as full of nutrients as animal manure. Use it as a fertilizer by spreading it along rows or working it into the soil of the entire vegetable garden before planting time. Two to four bushels of moist compost per 100 square feet are plenty.

As a Food Reservoir for Plants Place a shovelful of compost under hills of squash or cucumber before planting seed. Or scatter a thin layer along a furrow just before you sow peas. When roots begin the search for nutrients, the compost will be nearby and plants will grow by leaps and bounds.

OTHER WAYS TO IMPROVE YOUR SOIL STRUCTURE

Compost is by far the most popular and most effective of soil improvement methods. But there are other ways that deserve mention. Some will do things compost cannot.

Cover Crop A cover crop is winter rye or vetch planted in the fall. When the garden has been cleared of refuse and the soil has been loosened, the seed is simply scattered willy-nilly at the rate of three pounds per 1,000 square feet. Soon the roots take hold and the growth covers the soil preventing winter winds and spring rains from carrying off valuable topsoil. Then when the grass is turned under in the spring, with a spade, a plow, or a rotary tiller, the plant decomposes, adding nutrients directly to the soil. It's because of this fertilizer value that cover crops are often called green manures. To keep soil microbes from using up all available nitrogen, spread a high-nitrogen fertilizer over the garden area before plowing under the plants.

Mulching If soil structure is dangerously compacted and the surface often bakes into a hard crust, then mulching is definitely a must (see chapter 4). At the end of the growing season organic mulches can be worked directly into the soil, where they will break down in much the same way as in composting. At the same time, nutrients will seep directly into the garden soil.

Fall Plowing In areas where the soil compacts into rocklike clods it is worthwhile to till the soil before winter sets in. You'll gain a valuable ally in winter because of the alternate thawing and freezing, a natural process that breaks up clods nicely.

Manure Although the nutrient content of old rotted manure is somewhat low, it is one of the best materials that can be used in quantity to open up soil and give it a better structure. About 100 pounds or four bushels per 100 square feet of garden area should do the job. Fresh manure, which is higher in nutrient content, can also be used, but must be applied sparingly because of its "hotness"; that is, if applied directly to plants it can burn the roots. Play it safe by working fresh manure into the soil at least two weeks before planting time.

THE NUTRIENTS PLANTS NEED IN ORDER TO GROW

Besides providing a place for roots to grow, soil is a spongelike storehouse where crucial nutrients are kept close by. Not all soils have a proper balance, however. Some are naturally deficient; some have been depleted by being "overworked."

Fortunately, we're not helpless bystanders. There is plenty we can do by way of keeping the soil amply supplied with the nutrients plants need for rapid, healthy growth. But the market is glutted with "miracle" plant foods, commercially "prepared" fertilizers, organic materials, and liquid "instant growth" fertilizers. Which ones should be used? How much good will they do?

The best way to save energy, time, and money is to come straight to grips with what plants need in the way of nutrients. With a few basics in mind you can design your own fertilizing program, custom fitted to your particular soil. Let's take a close look at the major nutrients and what they do.

Out of a total of 16 elements needed by plants, 13 are found in soil. But some are more important than others. Nitrogen, phosphorus, and potassium are the top three and are known as major nutrients. Next come the secondary nutrients, calcium, magnesium, and sulfur. Finally, there are the micronutrients, iron, boron, manganese, copper, zinc, molybdenum, and chlorine. All of these make their way in solution with water from the soil to plant roots and ultimately into some part of the stem, leaf, fruit, or blossom. The remaining three elements—carbon, hydrogen, and oxygen—come from the air.

Of the 16 elements known to be necessary for plant growth, 13 come from the soil. The best friend a vegetable can have is a gardener who understands what the soil does and how it supports plant life. Only then can a fruitful soil-improvement program be undertaken and yields increased.

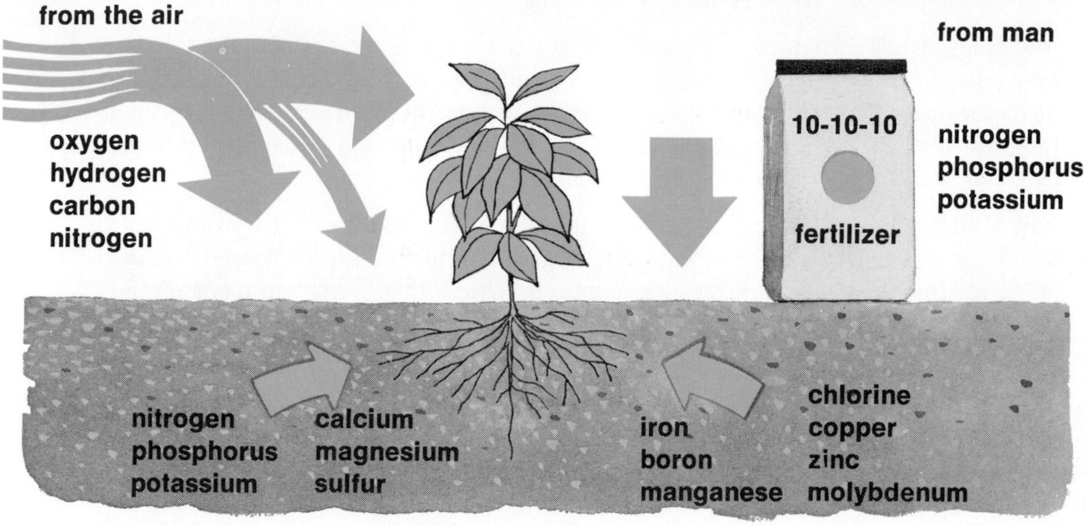

Nitrogen Means Healthy Leaves Anything that grows depends on nitrogen, and plants are no exception. A crucial building block for chlorophyl production and an indispensable element for cell respiration and reproduction, nitrogen promotes abundant leaf growth and is especially valuable for crops that are grown for their leaves, such as spinach, lettuce, or cabbage. Most of the nitrogen digested by plants is used during the early stages of growth. Without a sufficient amount of it, lower leaves will begin to yellow along the midrib and eventually will drop off, because nitrogen goes first to the uppermost leaves where growth is the most rapid. Too much nitrogen, however, will give all leaves and no fruit. Stems will become pulpy and weak, and finally the plant collapses without producing a yield.

The natural source for the nitrogen that eventually winds up in the soil is the atmosphere, some four-fifths of which is nitrogen gas. But plants cannot absorb this nitrogen directly—it has to go through a fixation process before it is available to roots. Snow, lightning, and rain supply some nitrogen to the soil, but the major natural suppliers are the millions of nitrogen-releasing bacteria living in the soil that absorb nitrogen from decomposing natural materials and release it for use by plants. Some legume plants such as peas have root nodules that take up nitrogen and release it after the fixation process for use by other plants.

Nitrogen can also be supplied to plants in artificial form by being chemically fixed in commercially prepared fertilizers. This makes it possible to control the amount of nitrogen available in the soil. It is unwise to overuse such fertilizer, however, for too much nitrogen will cause overly-leafy plants bearing little fruit.

Phosphorus Builds Healthy Roots and Flowers Without phosphorus the all-important process of photosynthesis, which converts water and carbon dioxide into carbohydrates, could not take place. Also concentrated in new growth, blossoms, and seeds, phosphorus has a lot to do with the arrangement of cells in plants. Fortunately, phosphorus doesn't disappear from the soil as rapidly as nitrogen, nor is it used in as great amounts. Yet plants will show hunger signs if there is a deficiency. Lower leaves begin to wither and turn yellow as if suffering from lack of nitrogen. The telltale difference is that a tinge of red or purple starts at the tip of the leaf and progresses along the outer edges. Another sign of phosphorus deficiency is a stunted root system.

Potassium Helps to Produce Disease Resistance and Strong Stems Potassium is a kind of jack-of-all-trades aide for the plant. It can help develop a resistance to drought or low temperatures, enhance the size, flavor, and color of vegetables and fruits, build strong stalks, increase the oil content in seeds or nuts, and promote resistance to disease. Root crops such as potatoes, carrots, beets, and turnips respond well to extra applications of potassium. When a plant is starving for potassium its older leaves will begin to turn yellow at the edges, growth will be stunted, and the stalk will be weak.

KEEP THE SOIL BALANCED

Resist the temptation to pour truckloads of fertilizer on your garden. All the nitrogen in the world will do nothing if any of the other elements are absent. The goal is to create a balanced soil in which all the nutrients are available to plants. By keeping an eye out for deficiency symptoms and by comparing growth from one season to the next, you can get a pretty good idea of the condition of your soil. Soil-testing kits for home use, while not as accurate as methods employed by experts, can also give a fair idea of soil deficiencies. If severe problems arise, send a sample of your soil to the nearest State Agricultural Experiment Station for an accurate analysis. The important thing is to preserve a balance of nutrients in the soil. A shortage of a single nutrient will hold back growth regardless of the quantity of other nutrients present. On the other hand, you may be fortunate enough to have land that is perfect for crops just the way it is. If so, it would be nothing short of folly to spend time and money on fertilizers that will only wash away after the first rain.

How can you tell? Watch your crops. If yields are good and plants are happy, then leave well enough alone. As a rule, new land being gardened for the first time will contain the necessary nutrients required by crops. Only after three or four years of consecutive cultivation will signs of depletion begin to appear.

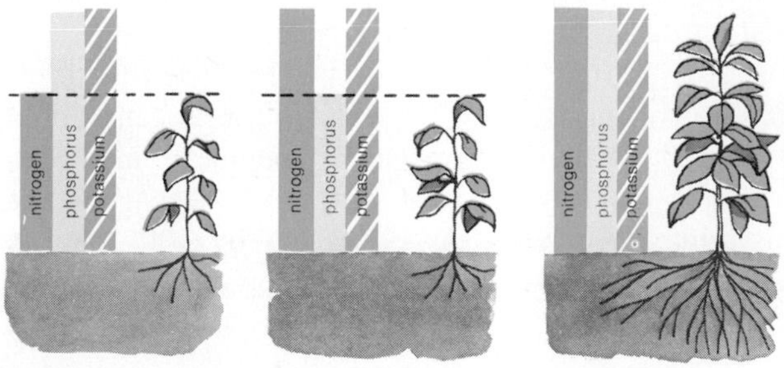

Plant growth will be limited by that element which is in shortest supply in the soil—thus the importance of having a *balanced* soil.

BRINGING YOUR SOIL UP TO SNUFF

If your soil needs a boost and you decide on a fertilizing program, you will be faced with making a choice—either chemical or organic fertilizers. Both have advantages and disadvantages. Here are some:

1. The nitrogen in most chemical fertilizers dissipates very rapidly. Plants will get a quick shot and look deep green for a short period, only to return to their original pallor. Slow-release materials designed to give off nutrients gradually after application are available, but they are considerably more expensive than regular fertilizers.

2. Organic substances such as dried blood, bone meal, sludge, and cottonseed meal take much longer to break down and as a result release nutrients slowly. At the same time they help to improve the physical make-up of the soil. The problem is that they are often difficult to come by, although a few telephone calls to local slaughterhouses, breweries, and other industries using raw organic materials will usually turn up a source.

3. Chemical fertilizers can be used immediately whereas fresh cow, horse, or chicken manure should have a chance to age before being applied to the garden. If used raw, they will "burn" roots. One way to avoid burning is by working manure into the soil at least two weeks before planting time to give the soil a chance to absorb the nutrients.

4. Chemical fertilizers have no effect on the physical condition of the soil. If drainage is poor or if the soil crusts, these fertilizers won't help.

CHEMICAL FERTILIZERS AND PLANT FOODS

Chemically "complete" fertilizers contain in synthetic, granular form the three basic elements needed for healthy plant growth. Safe and easy to handle, they offer the home gardener a chance for careful control of the amount of nutrients available to plants. If leafy crops are grown, then a complete fertilizer high in nitrogen (10–6–4) can be applied. If root crops are your specialty, a fertilizer with equal amounts of phosphorus and potassium but low in nitrogen (5–10–10) is the answer.

Whether you buy a liquid or granular fertilizer, the nitrogen, phosphorus, and potassium content is always clearly indicated by a numerical code. For example, a bag of fertilizer bearing the formula 5–10–10 contains 5 percent nitrogen, 10 percent phosphorus and 10 percent potassium. Frequently manufacturers will simply use the letters N, P, and K for nitrogen, phosphorus, and potassium. But the order of listing is always the same: first nitrogen, then phosphorus, and finally potassium.

Sometimes recommendations for the use of fertilizers are expressed in terms of weights of the actual elements, for example, 1 to 4 pounds of nitrogen should be used per 1,000 square feet of soil. To find out how many actual pounds of a nutrient the fertilizer contains, simply multiply the total weight of the bag by the figure representing that element printed on the label and divide by 100. Thus a 100-pound bag of a 5–10–10 complete fertilizer will contain 5 pounds of actual nitrogen, 10 pounds of actual phosphorus, and 10 pounds of actual potassium; an 80-pound bag of the same formula will contain 4 pounds of nitrogen, 8 pounds of phosphorus, and 8 pounds of potassium. The remainder consists of inert materials and trace elements.

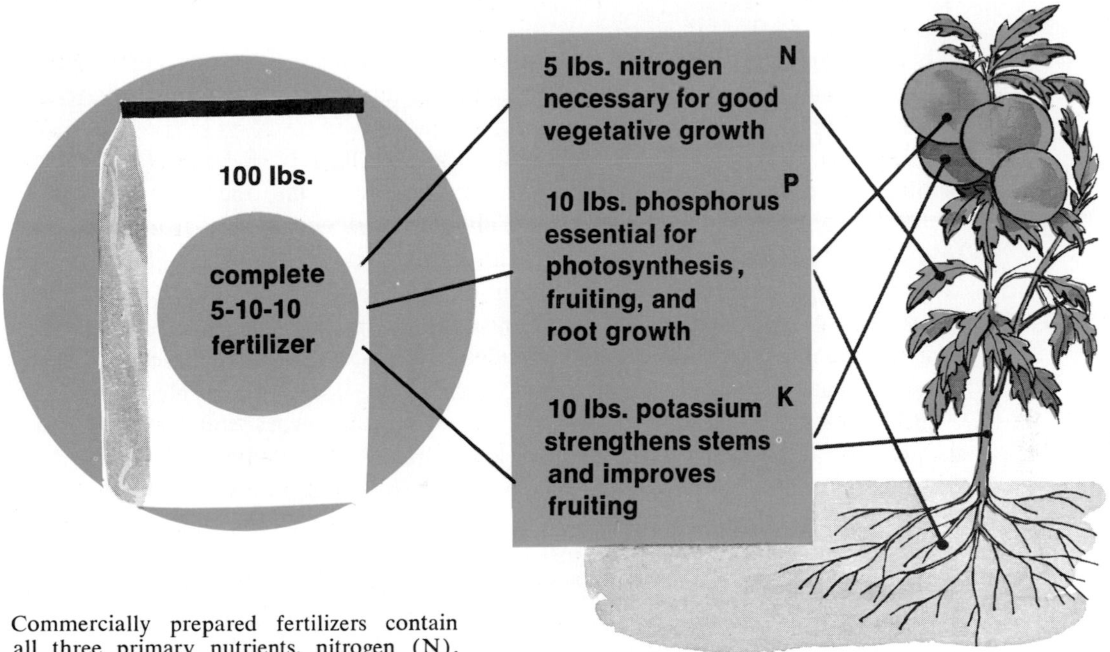

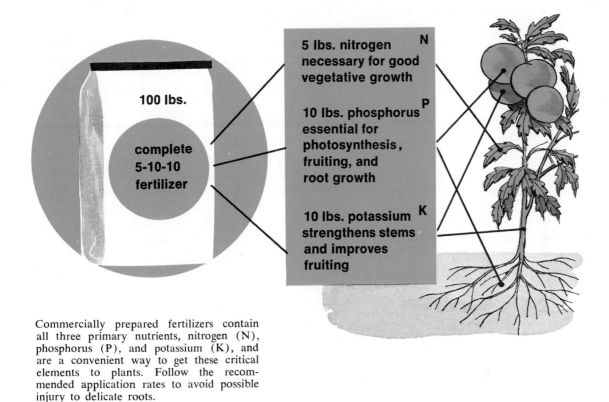

Commercially prepared fertilizers contain all three primary nutrients, nitrogen (N), phosphorus (P), and potassium (K), and are a convenient way to get these critical elements to plants. Follow the recommended application rates to avoid possible injury to delicate roots.

How Much to Apply Before applying any fertilizer, consider the crop you are growing and the type of soil. Leafy crops will improve with extra amounts of nitrogen, fruiting crops with phosphorus, and root crops with added amounts of potassium. The soil structure, too, will have an effect on the rate and frequency of application. A sandy soil allows moisture and nutrients to leach away rapidly, which means that doses of fertilizer, especially nitrogen, are needed at closer intervals than for normal soil. For compact soil with more clay content, applications should be somewhat lighter and not as frequently applied. To sum up: One application of a complete fertilizer on a compact soil may be sufficient for the entire growing season, while three or four applications per season may be necessary on a loose sandy soil.

For Average Soil (Loam)

Use 4 pounds
Fertilizer 5-10-10 (for most crops)
To cover every 100 square feet of garden area

If in doubt about drainage, try this simple test. Dig a few holes two to three feet deep at scattered points in the garden and fill them with water. If the water has disappeared within about 30 minutes to an hour, your soil has good drainage qualities. If, however, the water still remains after 24 hours, you can expect serious problems when growing crops and you should take steps to improve the drainage.

When to Apply Fertilizer As a general principle remember it's far better to fertilize in small amounts at more frequent intervals rather than to unload the entire season's supply in one application.

1. Start by bringing the garden area up to high fertility. Broadcast a complete fertilizer (see above) over the plowed area and work it into the top two or three inches of soil with a rake or harrow. This can be done anytime from one week to a day before planting.

2. Apply additional dressings after plants become established. Most plants make their greatest demands for nutrients during the early stages of development when you can almost hear stems and leaves growing. Sprinkle a light application of a high-nitrogen fertilizer (nitrate of soda or a 10–10–10 complete fertilizer) along both sides of the row, using about one pound for every 100 feet. Take care to prevent the material from coming into direct contact with stems or leaves. Chemical fertilizer, don't forget, is a concentrate and can seriously burn plants.

3. Give another similar dose when crops begin to blossom or set fruit. It isn't difficult to imagine the strain plants are under when everything begins to happen at once. The business of producing blossoms which will turn into fruit requires all the energy the plant can muster. You can help by making it a little easier for the roots to find the needed nutrients. Use a complete fertilizer and apply in the same way and at the same rate as described above.

ORGANIC FERTILIZERS

You might encounter difficulties in obtaining organic fertilizers, but their value as nutrient suppliers and soil conditioners far outweighs whatever inconvenience you experience in finding them. As a result of man's growing awareness of his environment, though, many organic substances are becoming more and more available. Try your garden supply center and, if that fails, a farmers' supply outlet. As a last resort you can always contact a slaughterhouse, winery, or sewage disposal plant in your locale.

Here are some common materials, their nutrient content, and the rate at which they should be applied.

Natural Sources for Nitrogen

	% NITROGEN	AMOUNT PER 100 SQ. FT.
Blood meal	15.0	10 oz.
Cottonseed meal	8.0	1¼ lbs.
Fish scraps	8.0	1¼ lbs.
Bone meal	4.0	2½ lbs.
Cocoa shells	2.7	3¾ lbs.

Natural Sources for Phosphorus

	% PHOSPHORUS	AMOUNT PER 100 SQ. FT.
Phosphate rock	30.0	¾ lb.
Bone meal	24.0	¾ lb.
Fish scraps	13.0	1¾ lbs.
Basic slag	8.0	2½ lbs.
Cottonseed meal	2.5	8 lbs.

Natural Sources for Potassium

	% POTASSIUM	AMOUNT PER 100 SQ. FT.
Wood ashes	8.0	14 oz.
Tobacco stems	7.0	1½ lbs.
Granite dust	5.0	2 lbs.
Seaweed	5.0	2 lbs.
Fish scraps	4.0	2½ lbs.

A quick glance at the listings will reveal certain materials that are rich in other elements as well. For example, fish scraps make an excellent fertilizer because they contain abundant amounts of all three substances—nitrogen, phosphorus, and potassium. Natural organic fertilizers are incorporated into the garden in the same manner as chemical fertilizers.

PLANTS AND THE PITFALLS OF pH

Suppose you have dumped wheelbarrow loads of fertilizer on your garden, patiently watered with the garden hose, and nipped every weed as it emerged. And still nothing happens. The problem could be simply a matter of pH, the technical term for the degree of acidity or alkalinity. All the fertilizer in the world won't help if the soil is too acid or too alkaline. Nutrients will remain locked up in the soil and unavailable to plant roots. Even worse, soil organisms can't thrive, and without these bacteria the nitrogen, phosphorus, potassium, and other nutrients become useless to plants.

Fortunately, you can change all this. By adding either lime for alkalinity or sulfur to acidify the soil, you can transform a barren garden into a forest of healthy plants. But some understanding of pH is essential. You can't be intelligent about adding lime or sulfur unless you know what you are doing and why.

What Exactly Is pH? As an example, we know that a lemon is acid, household baking soda is alkaline, and between the two is distilled water, which is "neutral." The whole question of acidity hinges on the number of hydrogen or hydroxyl ions present. A soil overstocked with hydrogen ions is said to be "acid" in reaction because this kind of soil makes iron, aluminum, and manganese so soluble that they will approach toxic levels for plants. At the other extreme, if too few hydrogen ions are present, the mineral elements become scarce or unavailable and plants starve. But plants differ in as many ways as people do. Some thrive in alkaline soils while others prefer acid soils—though most do best in soils that hover near the neutral mark. Vegetables and fruits are much more comfortable in soils that are neutral or slightly on the acid side.

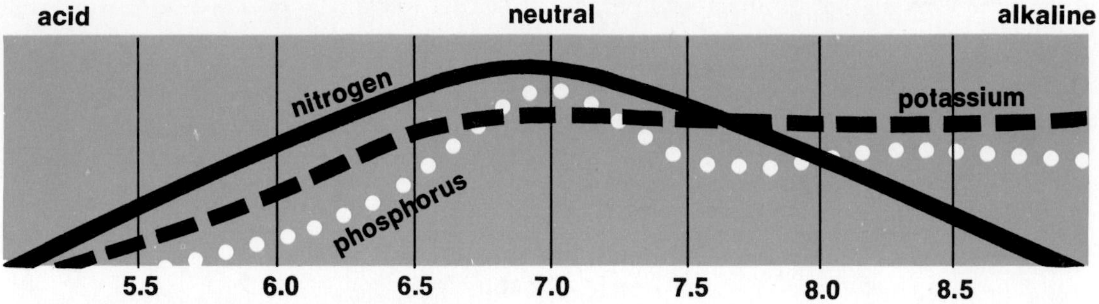

As the pH reaction of the soil approaches 7.0 (neutral), nitrogen, phosphorus, and potassium reach their peak of availability to most plants.

The pH Scale Don't let the mathematical juggling of pH send you or your gardening plans into a hopeless tailspin. All you need remember is that the pH reaction of a soil is expressed on a simple scale ranging from 0 to 14. If a soil sample tested at home or in a laboratory registers toward the 0 end of the scale it is said to be acid; it is alkaline if it climbs toward the 14 mark. Soils that test near the middle (7.0 on the scale) are either neutral, slightly acid, or slightly alkaline. Most soils fall between pH 4 and 8. But vegetables and fruits need soil that is between 5.7 and 6.9 in pH reaction. In other words slightly acid conditions are best.

How to Determine pH You can get an idea of the pH level of your soil by observing which plants thrive there. (See following table.) A more accurate method is to test soil samples with a home testing kit or, best of all, by packing off samples to your State Agricultural Experiment Station. Although this requires a little more effort and perhaps a small fee, samples sent to state agricultural centers will be evaluated by experts using expensive and sensitive equipment. If deficiencies exist they will tell you what they are and what corrective measures to take.

Judging pH by Observation A lot can be learned about your soil by simply observing what grows there and the kind of climate you have. Although hardly as accurate as scientific analysis, watching certain acid- and alkaline-sensitive plants can tell you something about the soil's pH reaction. Beets, for example, are not tolerant of acid conditions and will do poorly. Potatoes, on the other hand, will suffer from scab if not planted in a fairly acid soil.

Climate and soil conditions can also have a bearing on the degree of acidity of your soil. Some other indicators:

A sandy or clayey soil is likely to be more acid than a loam soil.

Pine, oak, azalea, blueberry, raspberry, and rhododendron all prefer acid conditions. If these varieties are prevalent in your area, you might want to add a bit of alkaline "sweetener" to your vegetable garden plot.

Areas where rainfall is moderate to heavy are likely to be on the acid side because of the chemical changes that take place when rainwater percolates through the soil (see map on page 52). Thus soil in the eastern half of the U.S. is generally more acid because of heavier rainfall.

pH Preferences of Common Vegetables and Fruits

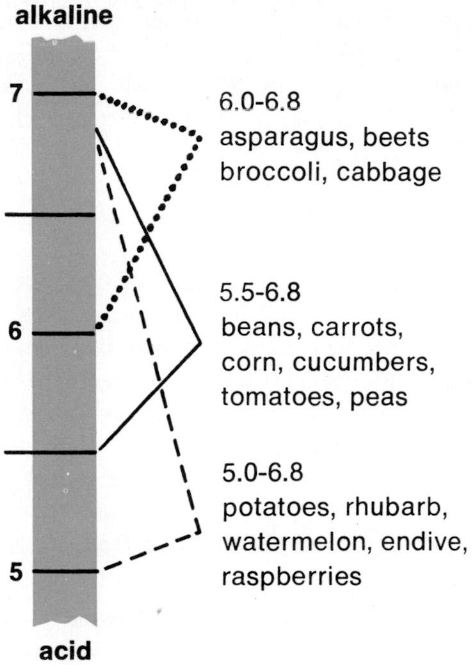

Testing Your Soil at Home Home testing kits, available from seed catalogs and local garden supply centers, can give you a workable picture of how your soil fares on the pH scale. In addition, soil samples can be tested for nitrogen, phosphorus, and potassium levels. Care must be taken to obtain several samples to a depth of at least six inches. Follow the manufacturer's directions carefully.

State Agricultural Experiment Station Tests By far the most accurate, these tests by the experts will show exactly what the pH is, what the soil needs, and in what quantities. When sending off a sample, be sure the container is labeled to provide information concerning the location and depth of the sample, what previous fertilizers have been used (if known), and the crops to be grown.

WHAT TO DO ABOUT pH

Once the pH level is known, improving the soil is as simple as applying a balanced fertilizer. But the material you use depends on the pH you wish to achieve. For an acid soil the most useful "sweetener" is either ground dolomitic limestone or agricultural lime, both available at any farmers' outlet or garden supply center. When the soil is too alkaline, powdered sulfur will bring the pH to a level at which crops will produce high yields (see chart below). Before applying either of these elements, keep the following points in mind:

1. Ground dolomitic limestone and powdered sulfur are concentrated materials that can do untold injury to plants if used carelessly. Stick to the recommended rates of application.

2. Even if your soil calls for drastic measures, limit applications to one at the beginning of the growing season and another at the end. Make sure lime is thoroughly worked into the soil.

3. Before calculating lime applications be sure they coincide with the pH preferences of the crops you intend to grow. Refer to the chart on page 33 for specific crops and preferred pH levels.

4. Overapplication of lime or sulfur can cause the very same nutrient problems you are trying to correct. Go easy and apply them a little at a time.

Whatever the results of a soil test may be, don't feel as though you should hurl bagfuls of lime or sulfur into every corner of your garden. Plants have an amazing ability to make do with what they have. Vegetables and fruits, in particular, will surprise you with their hardiness and willingness to grow almost anywhere. Even potatoes, which are susceptible to potato scab disease when grown for several years in neutral soil, will probably do fine if the crop is rotated to another part of the garden each year. Sometimes the addition of plenty of organic matter will cure excess soil acidity, depending on the components added.

To Change pH in Your Soil

To raise pH from 4.5 to 5.5 use 60 lbs. ground dolomitic limestone for every 1,000 sq. ft. of garden area (6 lbs. per 100 sq. ft.).

To raise pH from 5.5 to 6.5 use 85 lbs. ground dolomitic limestone for every 1,000 sq. ft. of garden area (8½ lbs. per 100 sq. ft.).

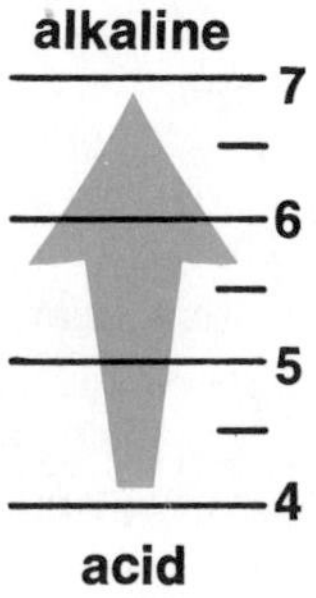

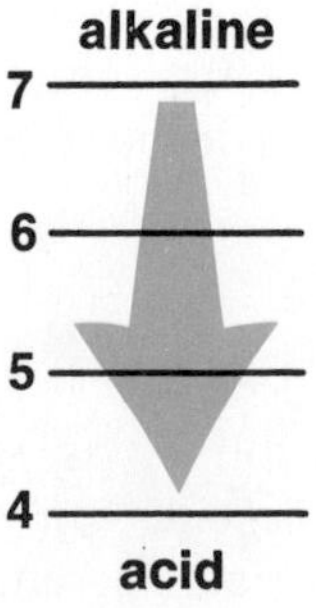

To lower pH from 7.0 to 6.0 use 4 pts. powdered sulfur for every 100 sq. ft. of garden area.

To lower pH from 6.0 to 5.0 use 4 pts. powdered sulfur for every 100 sq. ft. of garden area.

3.

Planting Seeds - Indoors and Out

In the gardening world one vegetable's "meat" can easily be another's poison. For example, some vegetables, such as peas, thrive on cooler temperatures while others need warm days and nights. Because of these differences, specific instructions for growing each crop are given in chapter 6 and 7. But there are a few basic principles that might help you keep things green and growing when setting seeds to soil.

OUTDOOR PLANTING

Many seeds can be sown directly outdoors, but you will have to work within the limits of the climate, the weather, and the site of the garden. If you are lucky enough to possess a "quick" soil that drains well and warms up rapidly, you can get under way as soon as spring has officially arrived. (See appendix on page 218 for frost-free dates in your area. The map on page 36 gives a general idea.)

Preparing the Soil The first step is to get the soil into proper condition for receiving seeds. Timing here is important, since the soil must not be too wet. Test the soil first by squeezing a small amount in your hand. If it crumbles easily then the moisture level is just right and the soil will be a joy to work with. If not, wait until it is. Using a spading fork, a rotary tiller, or a standard plow, turn over the upper six to eight inches of topsoil. Break down clods with a harrow or hoe. Then rake to remove stones, twigs, and other debris that might obstruct crop growth. When ready, the soil should be loose and covered with a level top layer of fine soil in which seeds can germinate under ideal conditions.

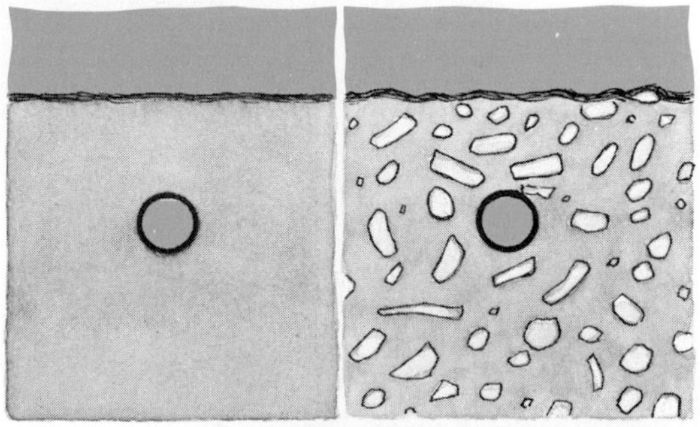

Seed planted in a well-prepared soil will germinate quickly. Seed planted in a rough soil littered with sticks and stones may not sprout at all.

Average Last Day of Frost in Spring

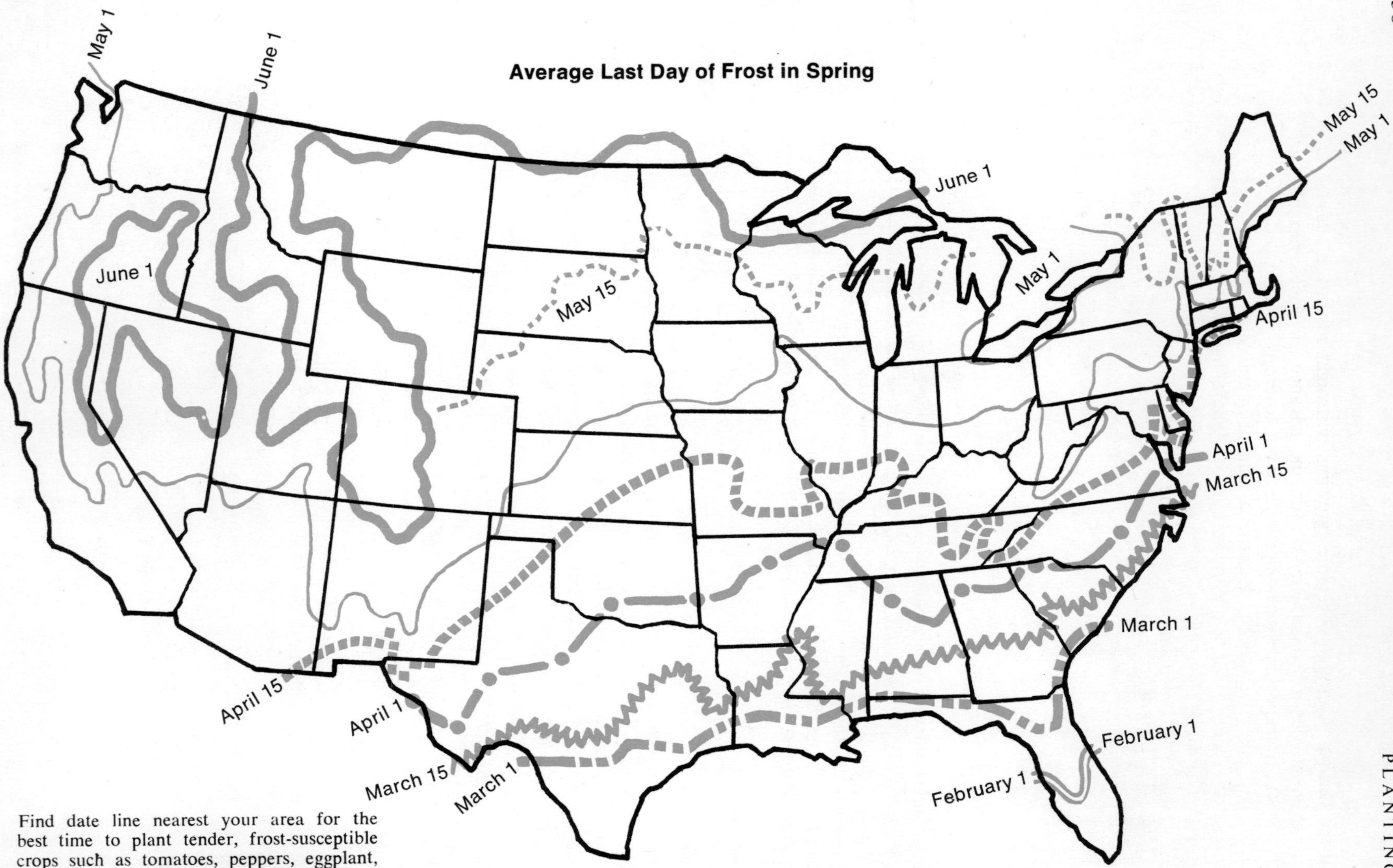

Find date line nearest your area for the best time to plant tender, frost-susceptible crops such as tomatoes, peppers, eggplant, and corn. For exact dates see the appendix at the end of this book, or consult local growers or your State Experiment Station. Dates will vary depending on geographical peculiarities of your specific region.

Laying Out the Garden Although vegetables growing in straight rows won't taste any better than vegetables growing helter-skelter, the advance of weeds can be checked more easily when a semblance of order is given to crops. Hand weeding, hoeing, cultivation, and other chores become a simpler matter when rows are straight and evenly spaced. Get a general picture of where the different crops are to go by placing stakes at both ends of each planned row. Keep one eye glued to the garden plan you prepared during the winter (see page 11). If possible place rows so they will run north and south unless the slope dictates east to west (or other) rows to prevent soil erosion. Remember that taller crops can keep smaller crops in constant shadow if not kept to the north side of the garden.

To keep rows straight, even, and easy to protect from marauding weeds, use heavy twine stretched between two stakes as a guide when preparing furrows.

Setting Seeds to Soil Once each vegetable has been assigned a section of the garden, you can get down to the exciting business of sowing the seed. Use this procedure for each row: First tie garden twine to the stakes at the ends of the row. Then dig a shallow furrow with a hoe, using the taut string as a guide. The best way to plant seed is by placing a small amount of seed from the package in the palm of one hand. Take a pinch of seed between the thumb and forefinger of the other hand and drop it into the furrow by rubbing fingers together. After two or three seasons of practice your rows will be perfectly spaced. Larger seeds such as beans or beets can be individually placed. The smaller seeds can be sown on top of the soil and gently rubbed into the ground by brushing your hand over them. The proper depth is, of course, crucial. As a general rule, plant seeds to a depth equal to four times their diameter. After seeds have been sown along the entire row, cover with a thin layer of soil. To insure proper moisture and darkness, tamp the soil *gently* with your hand or the back of the hoe. Sprinkle with a fine spray from the garden hose. Remove stakes and garden twine, and label the row.

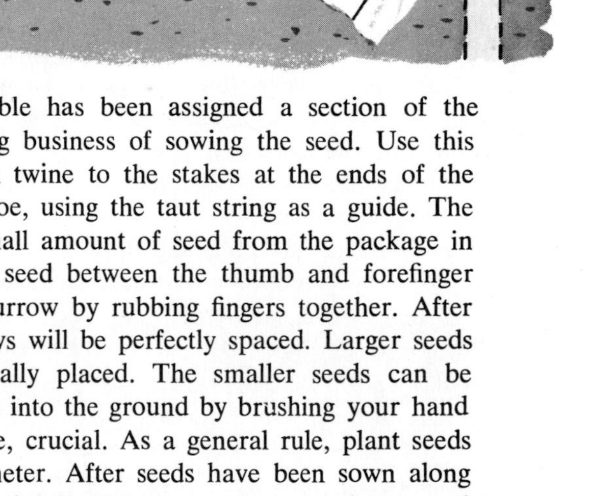

Large seeds are planted deeper in soil than small seeds. Depth is determined by seed's diameter—usually, four times the diameter.

Thinning Because more seeds than necessary are planted to allow for those that won't germinate, the row will always be somewhat crowded when seedlings emerge. If left to their own devices, the young plants will crowd each other out until the common denominator is a weak, spindly plant that yields little if anything. Therefore you should thin the row as soon as the seedlings appear established. For the first thinning, pull out every other plant. After a week or two, or when plants seem to be cramped, thin again. A row may require up to four thinnings before the plants reach full size. The later thinnings can be taken to the dinner table. As a guide, remember the leaves of plants may touch each other but should not overlap.

Tips for Planting Seeds Try to keep these facts in mind when sowing seed:

1. It is easy to overplant. Seeds sown too thickly will result in a matted row that is difficult to thin.

2. The deeper you dig into the soil, the cooler it gets. In order to insure the proper temperature for good germination, it is better to plant seeds on the shallow side in spring rather than too deeply. Late plantings or those in warm areas may go deeper.

3. Some seeds need a little pampering before germination will take place. Celery and parsley, for example, need an overnight soaking in tepid water to soften their hard shells. Others such as beets don't have the strength to push through crusted soil and should be covered with a thin layer of peat moss mixed with sand.

4. A row containing slow-germinating seeds may be engulfed in weeds before the seedlings finally appear. To mark these rows, mix some radish seeds, which germinate rapidly, with the other seed before planting. By the time these clearly visible rows have been made free of weeds, the radishes will have matured, the other seeds will have germinated, and the radishes can be pulled and taken to the dinner table.

5. Each vegetable and fruit is offered in a number of different varieties. The variety you choose will depend on how you plan to use the crop and where you live. There are three ways to find the best variety for your area. (1) Study the seed catalog carefully for descriptions that give clues as to where particular seeds will grow the best. (2) Buy seed from a farmers' supply center. Local farmers, unwilling to risk losing several acres of crops, have probably already determined the best variety for their soil as well as for their market. Chances are your soil conditions are similar to theirs. (3) For expert opinion get in touch with the nearest State Agricultural Experiment Station, where scientists are continually searching for bigger or better varieties adapted to local climates and conditions.

INDOOR PLANTING

The delight that comes from watching tiny seedlings growing in the dead of winter is one aspect of gardening no family should miss. Children will be practically mesmerized by the pebblelike seeds that burst into life before their eyes. For the gardener, the reward will be a longer growing season. Starting seeds indoors can add as much as three months to your vegetable-growing summer. Also, seeds are small and make very simple demands. All you need are some home-made containers, starting soil, a handful of seed, and enough space in a sunny window—requirements that can easily be met in any home.

THE FOUR PHASES

Because conditions indoors are not to be compared to the uncertain twists the weather outside can take, plants started inside must undergo four steps to prepare them for an outdoor life:

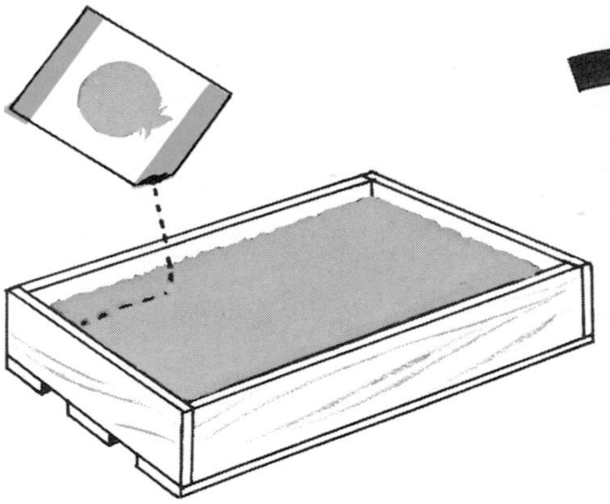

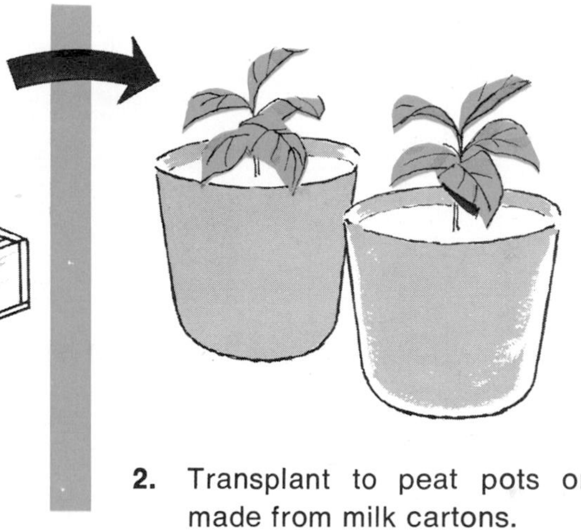

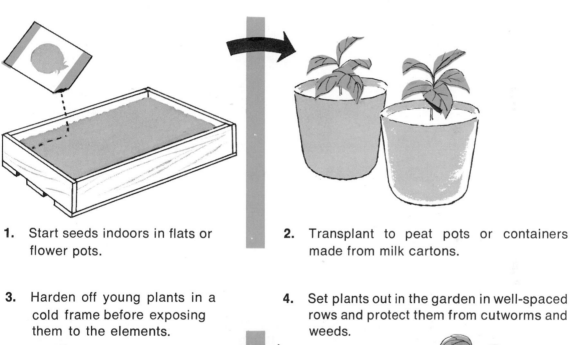

1. Start seeds indoors in flats or flower pots.

2. Transplant to peat pots or containers made from milk cartons.

3. Harden off young plants in a cold frame before exposing them to the elements.

4. Set plants out in the garden in well-spaced rows and protect them from cutworms and weeds.

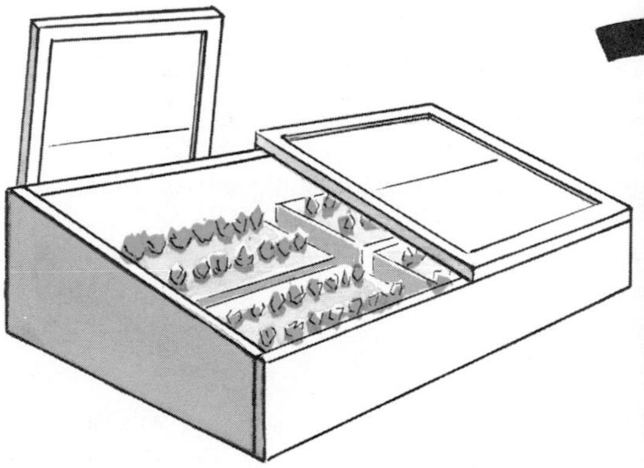

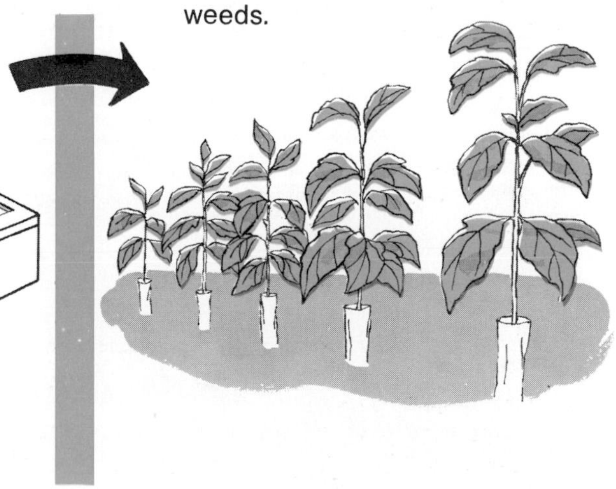

Initial Planting There are a number of ways to start seeds indoors, but by far the most popular is to use flats constructed of wood or plastic and filled with several inches of starting soil. A bag of house-plant potting soil can be purchased, or you can make your own starting soil by mixing together equal parts of garden loam, peat moss, and sand. Spread at least two inches of soil over the bottom of the flat and top with a sterile half-inch layer of vermiculite or sphagnum moss to prevent "damping off" fungus from killing the tender sprouts. Give the flat a thorough soaking with water. Then, after excess water has drained off, mark rows four inches apart and plant the seed. A piece of glass is a perfect cover for the flat to keep moisture in the soil. Since seeds need darkness in order to germinate, the flat should be put in a dark place or covered with newspaper. Keep track of varieties by inserting labeling stakes at the row ends on one side of the flat. When seedlings have emerged, thin them and move the flat to a sunny window. Turn the flat at least every other day to prevent sun-seeking plants from growing permanently crooked.

Flats for Indoor Planting

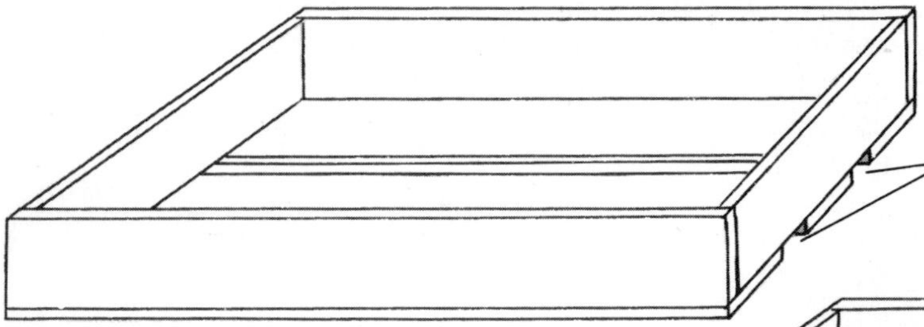

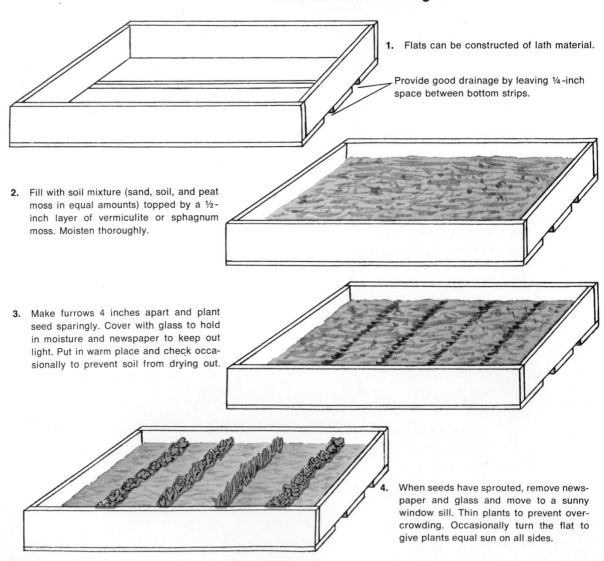

1. Flats can be constructed of lath material.

Provide good drainage by leaving ¼-inch space between bottom strips.

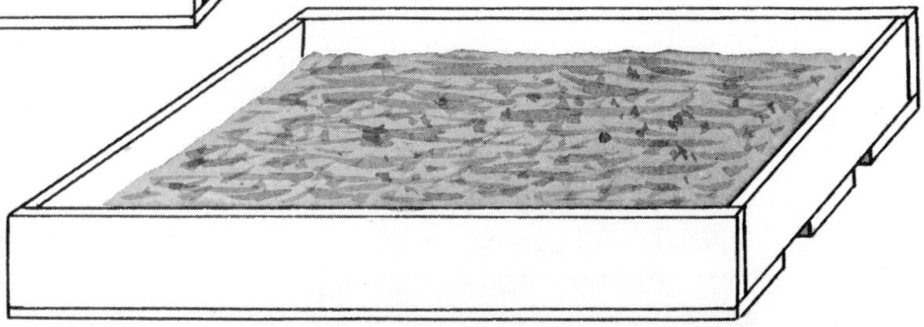

2. Fill with soil mixture (sand, soil, and peat moss in equal amounts) topped by a ½-inch layer of vermiculite or sphagnum moss. Moisten thoroughly.

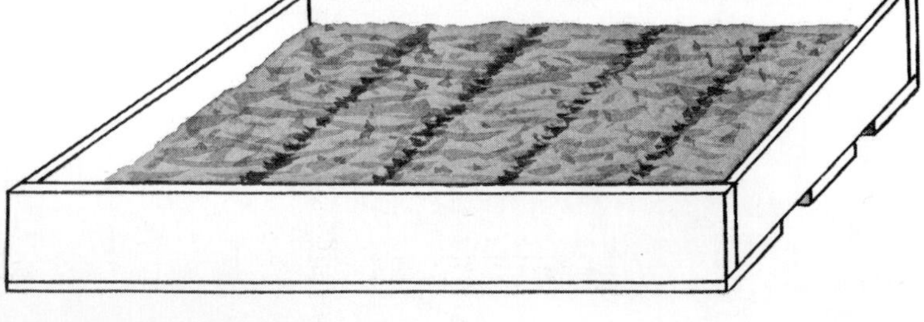

3. Make furrows 4 inches apart and plant seed sparingly. Cover with glass to hold in moisture and newspaper to keep out light. Put in warm place and check occasionally to prevent soil from drying out.

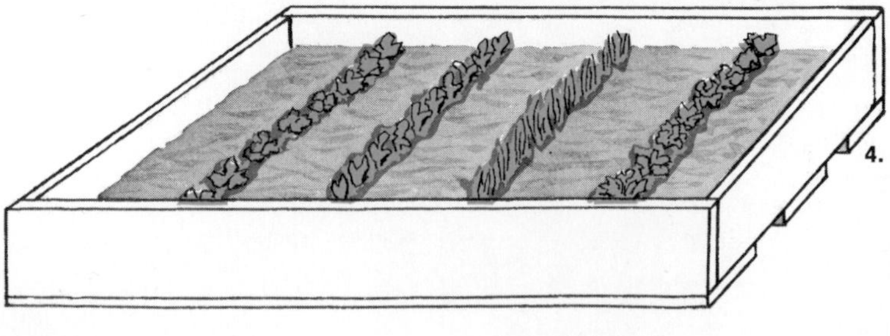

4. When seeds have sprouted, remove newspaper and glass and move to a sunny window sill. Thin plants to prevent overcrowding. Occasionally turn the flat to give plants equal sun on all sides.

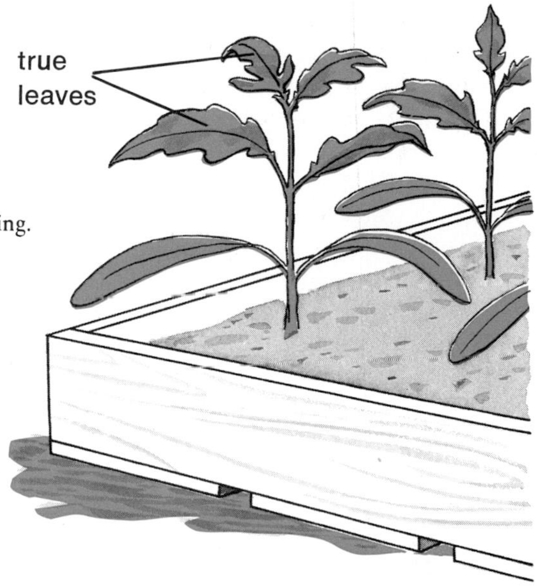

This young plant is ready for transplanting. Note true leaves above seedling leaves.

Transplanting When the young plants have developed two sets of true leaves (see illustration) they are ready for transplanting to small individual containers where they will remain until set out in the garden. Water the flat thoroughly so soil sticks to the tender roots. With a discarded kitchen knife or a notched stick lift out each plant along with a generous amount of soil and replant it in an individual peat pot or cardboard container, adding soil as necessary to fill in. Watering immediately with a weak solution of liquid fertilizer will help plants through the shock of transplanting. Then return the plants to the window sill. It's best to handle plants by their leaves during this transplanting process because the brittle stems are very susceptible to damage. The individual pots should still be turned every few days and checked regularly for proper moisture content.

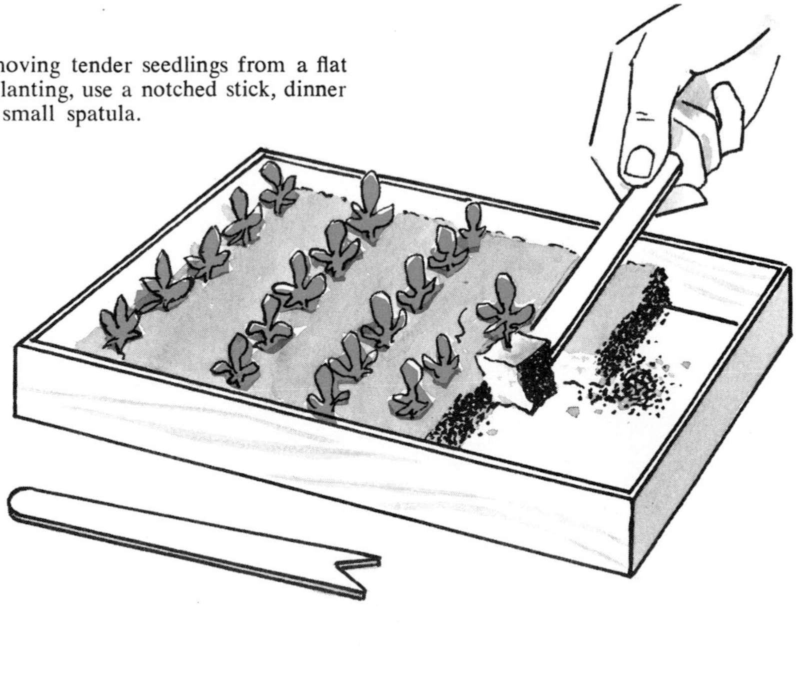

When removing tender seedlings from a flat for transplanting, use a notched stick, dinner knife, or small spatula.

Hardening Off Few young plants, regardless of how tough they appear, can stand the abrupt change from an indoor existence to an outdoor existence. To strengthen plants and prepare them for cold nights and windy days outdoors, they should be kept in the pots and spend a couple of weeks in a cold frame. Don't let the term "cold frame" frighten you—it need be nothing more than a box with a glass top placed outside where it gets plenty of sun.

If you can find one or two discarded storm windows, then your cold frame is already half complete. Simply measure the dimensions of the windows and build a wooden box to fit. Give the sides a slight slant (see illustration) so that when the cold frame is placed facing south the plants will get the maximum amount of sunshine. Figure a six-inch difference between the back, which should be at least 18 inches high, and the front, which should be at least 12 inches high. Place in a sunny location convenient to the garden with the slope facing south. Soil banked against the north side will help to keep cold winds out and warmth in. Fashion a notched piece of wood to prop open windows for varying the ventilation (see illustration) and your portable cold frame is ready for use.

Since the cold frame is not thermostatically controlled, it is up to you to make certain that plants get adequate ventilation and even temperatures. Remember, although chilly winds linger well into spring, the sun is much higher in the sky in spring and summer and can generate considerable heat in protected places like the cold frame. Keep an eye on your plants. On warm days prop the windows open or remove them entirely. Water plants when the soil appears dry.

A cold frame is not just a temporary device used only to capture the strong sun of spring. When plants have been moved to the garden, you can use the protected area in the cold frame for early starting of seeds for fall crops, such as cabbage, broccoli, or parsnips. In the fall, the cold frame can supply lettuce and other crops way beyond the first killing frost. Other vegetables, especially root crops, will keep through the winter if stored in the cold frame and covered with an insulating layer of hay or other mulching material.

The Cold Frame—A Plant's Halfway House

A cold frame is an excellent place to prepare plants that have been started indoors for unpredictable outdoor conditions. Consisting of a wood frame covered with a window sash, a cold frame can be put together in less than an hour's time. A block props up the sash for daytime ventilation.

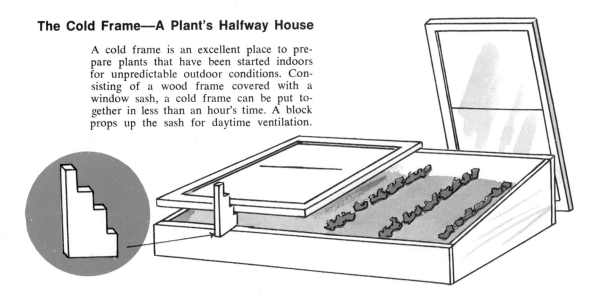

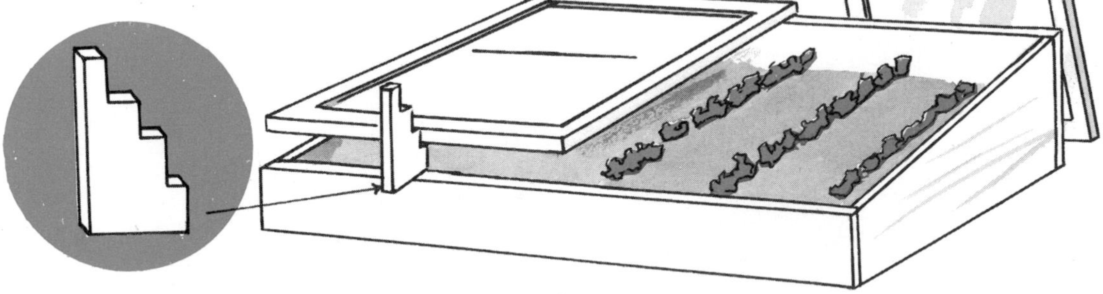

Setting Plants Out in the Garden As soon as all threat of frost has passed, the hardened-off plants can be placed in the garden. But unless care is exercised, this final move can be fatal. Follow these simple steps in order to give plants every possible advantage.

1. Wait for cloudy weather. Excessive heat can do in a transplant quicker than almost anything else, so choose a gray day for transplanting. If this is impossible, wait until late afternoon when temperatures begin to decline slightly. If the next day turns out to be a scorcher, give the plants a break from the intense sun by shading them with bushel baskets, newspaper cones, cheesecloth, or protectors manufactured commercially.

2. Keep plants protected during their travels from the cold frame to the final planting. Roots left out of their containers and exposed to the sun and wind will dry out in no time. Be sure to give plants a thorough watering beforehand, and plant only two or three at a time.

3. Mark the row with garden twine and dig evenly spaced holes according to the recommended spacings indicated in chapters 6 and 7. Fill the holes with water in order to saturate the soil with enough moisture to hold the plants over the first few days.

4. Set out plants as soon as the water has soaked into the surrounding soil. If peat pots have been used, simply place the plant, pot and all, into the ground. The pressed peat of the pot will gradually disintegrate as roots push out. Plants removed from other containers must be handled carefully so that the root systems nestle properly in the soil. Set plants so that the new soil line is somewhat higher on the stem than the old soil line. Then fill in the hole and press the soil gently to firm it around the roots. But don't press too hard. Overcompacted soil can seriously pinch roots and the plant will be long in recovering if it recovers at all. As a final step, leave a slight depression at the base of the plant to collect all available moisture.

Do your transplanting on a cloudy day if possible and work quickly to prevent fragile roots from drying out. Water the area before and after planting.

Plants produce roots according to the demands of the leaves. In transplanting, many of the finer root hairs are injured. Snipping leaves reduces the call for nutrients and prevents overtaxing of recuperating roots.

Planting and Transplanting

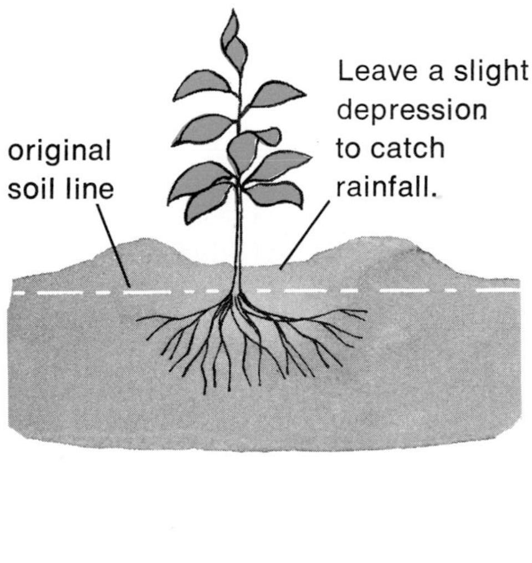

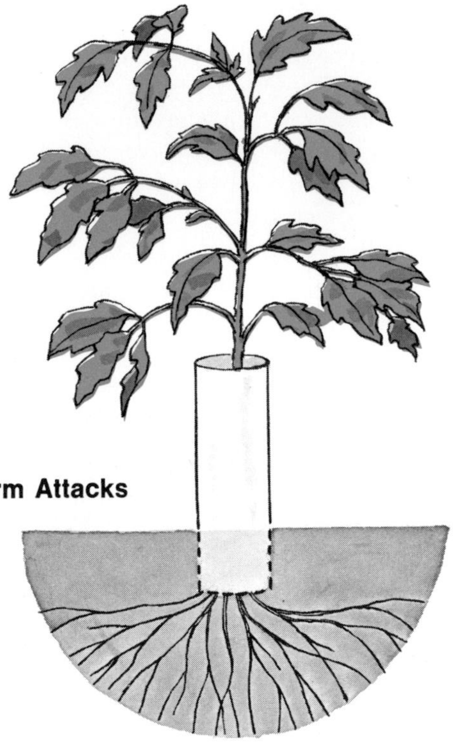

Protection Against Cutworm Attacks

5. Provide protection against cutworms for cabbage, eggplant, peppers, and tomatoes by placing a three-inch collar of aluminum foil, stiff paper, or cardboard around the stem of each plant and inserting it one inch in the soil.

6. Water garden area thoroughly and at least every other day thereafter for the first week or so or until plants are well established.

Understandably, many gardeners do not have the time, the facilities, or the patience to embark on this four-stage seed-starting process. But long-season crops need not be abandoned simply because of lack of time. Most nurseries, greenhouses, and garden supply centers are bursting with prestarted plants of all kinds in flats by the time the spring season arrives. Well fed and healthy, these purchased plants have already been hardened off and can be placed directly in the garden from their flats or containers as soon as all danger of frost is past.

4.

Weeds and Water

Just because the seeds have been sown doesn't mean your gardening responsibilities are over. There is the inescapable march of weeds and the ever present threat of a drought. But one of the secrets of successful gardening involves small doses at regular intervals. Little things mean a lot. It is far better to do 20 minutes of gardening every day than to save all the chores for one whole Saturday each month. The reasons are simple. To begin with, a small weed is easier to dislodge than a big one, which will probably take several vegetable seedlings along with it. Bugs nipped in the scouting stage don't have a chance to become numerous enough to do permanent damage. And plants given treatment as soon as their leaves look wilted will recover rapidly. In short, a plant's best friend is the gardener's watchful eye, its worst enemy a procrastinator who puts off for tomorrow what should have been done yesterday.

Cultivation The main purpose of cultivation is to thwart weeds before they have a chance to block the sun, prevent good air movement, soak up valuable soil nutrients, and harbor injurious insects. But frequent shallow working of the soil can have other advantages:

Soil crusting is prevented.

Air can penetrate a loosened soil more easily.

Grubs and pupae of undesirable insects are exposed.

A crumbly soil will soak up rainfall and prevent erosion from water runoff.

Try to get the jump on weeds, especially when vegetables and fruits are still tender and unable to fend for themselves. Later on, when they mature, their leaves will cast a dense shade and make it more difficult for weeds to take hold. The first few weeks are critical. Keep a constant vigil and remove weed invaders as soon as they are visible.

The only way to rid the row itself of emerging weeds is by hand pulling. Weeds that appear between the rows can be quickly dispatched with a gentle sweeping motion of a sharp hoe. Remember to keep cultivation shallow, especially close to the row where plant roots are likely to be near the surface. Don't forget the weeds that crowd the garden's edges. Keep them cut with a hand sickle or a rotary lawnmower, to remove a hiding place for insects and to improve the appearance of the garden.

A misused hoe can do more harm than good. The hoe on the left penetrates too deeply, will cut and tear shallow root systems. The sweeping motion on the right is correct, will nip sprouting weeds and aerate soil without threatening roots.

THE MIRACLE OF MULCH

Nature's system of keeping the soil healthy is truly remarkable. Dead leaves, twigs, stalks, and other vegetable matter fall to the ground and form a natural layer of plant material we call a *mulch*. This gradually breaks down through the action of soil-borne bacteria and decomposes, refurbishing the soil. For the home gardener, it is a process worth imitating. Here's what a good mulch can do:

Smother Weeds Like anything else, weeds need sunlight to grow. They will languish and die beneath a good mulch. The few weeds that manage to sprout through it will be haggard and weak, easy to remove.

Preserve Moisture in the Soil A loose mulch will minimize the drying-out effects of a strong sun and brisk winds. Ordinary evaporation is also held down so that moisture stays closer to the roots of plants.

Add Nutrients Directly to the Soil In terms of providing nutrients, mulch scores higher than compost because the material decomposes right in the soil rather than in an isolated spot away from the garden.

Improve Soil Structure Once an organic mulch has been turned under and incorporated into the soil, the moisture and nutrient-retention qualities of the soil will be much improved. Your soil will also become friable and a joy to work with.

Hold Down Soil Temperature Because mulch blocks the direct rays of the sun, the soil will not heat up as rapidly. Baking and crusting as well as root injury will be prevented.

Keep Vegetables Clean and Insulated from Soil-borne Diseases A soft layer of mulch not only will prevent soil runoff from sudden rains but also will keep raindrops from splashing mud on the vegetables and fruits. They are also kept from direct contact with the soil where diseases are often harbored.

Encourage Worms One of the best soil builders around is the ordinary earthworm. Because of its constant burrowing, the worm aerates the soil and creates pore spaces where moisture can be stored. Mulches provide ideal conditions for earthworms.

Impact of rain can splash soil, resulting in soil erosion and spattered vegetables (left). A good mulch (right) will prevent this.

Almost anything decomposable will do as a mulch, providing it is fine enough to form a plush layer around plants. The important things are depth and openness. A thin mulch will have little effect on weeds, while a mulch that is too heavy and compact can cause soil ventilation problems. A six-to-eight-inch depth is about right. Once settled and compacted, however, this should be no more than two to three inches thick.

Begin your mulching program right away. As soon as seeds have sprouted and seedlings are about four inches high, spread mulching material between plants and between rows. As the season progresses and the mulch breaks down, add fresh material to preserve the proper thickness. Then, in the fall, turn everything under to complete the decomposition process and improve the soil. But remember there are two sides to every coin. Mulch is not a cure-all. Some dry materials such as pine needles can become a fire hazard, while others such as soft leaves and peat moss can become so compacted that the soil remains cold and poorly ventilated, a perfect haven for diseases.

The cheapest source of mulching materials is your own backyard, where leaves and lawn clippings are always in good supply. If your yard is not large enough to meet the mulch demand, visit industries that use organic matter, such as wineries, sawmills, and candy factories (cocoa shells). As a last resort, many suitable materials are available from garden supply centers, but by the time you pay for fancy packaging you may feel a long way from imitating nature.

Here are some other good mulching materials, their characteristics, and some hints on the best way to use them.

HAY—Hay can be easily obtained from farms or from mowings on your own property. Sometimes farmers are more than willing to dispose of "spoiled hay" which has become too wet and moldy for livestock. It makes a perfect mulch and is especially useful around fruit bushes and trees.

LAWN CLIPPINGS—Since most people have some semblance of a lawn, lawn clippings might well be the closest thing to a universally available mulch. But they must be handled with some care. Green clippings can compact and cause excessive heating. Let the clippings dry for a day or two before spreading on the garden.

LEAVES—Leaves make an excellent mulch and are free for the gathering. A more manageable material is produced if you first set leaves aside and allow them to break down somewhat through composting. Maple and other soft leaves tend to compact and may, when wet, harbor disease spores. Mix such leaves with other materials. Oak leaves are tougher and may need crushing or shredding. They also may acidify the soil a little.

STRAWY MANURE—Manure is gardener's gold, but don't be too hasty about spreading it fresh from the stable. New manure can burn plants. When mixed with straw, manure makes an excellent mulch.

PEAT MOSS—Peat moss is pleasing to the eye and possesses good soil-building qualities. It is readily available at garden supply centers and is easy to handle. Unfortunately, it sometimes dries out, packs down, becomes crusted, and is finally impervious to water. Mix it with sawdust or wood chips for best results.

PINE NEEDLES—Pine needles make a perfect protective cover for the strawberry bed because they form a uniform layer and are not too easily dislodged by the wind. Find a pine grove and simply collect the needles that have fallen to the ground. They are, however, slightly acid in reaction. On soils with a pH of 4.0 to 5.6 a little liming may be needed.

SAWDUST—As an all-around mulch, few materials are as versatile as sawdust. Available from lumberyards, sawmills, or furniture manufacturers, sawdust looks nice, is easy to handle and cheap. However, don't let sawdust rob the soil of available nitrogen—add a nitrogen fertilizer or blood meal each spring or when plants look a paler green than you think they should.

WOOD CHIPS—Often found where sawdust is obtained, wood chips make a handsome mulching material. They are also available from road crews, street departments, and utility companies that prune around their wires and chip up the prunings. Use chips around raspberries or blueberry plants.

PLASTIC FILM—Plastic, of course, is hardly an organic substance but it is useful as a mulch. It is easy to use, practical for large areas, and unsurpassed as a moisture-preserving device. Both black plastic and clear plastic are available at garden centers and hardware stores. Punch holes here and there to let water get through into the soil.

Remember the Bacteria When a mulch or any other organic material is incorporated into the soil, thousands of nitrogen-consuming bacteria spring into action. Without nitrogen, these organisms starve and the breakdown process is slowed. To keep decomposition going at high speed, spread a complete fertilizer over the garden in the spring and turn both fertilizer and mulch under together each autumn. Do the same when a cover crop or over-the-winter mulch is turned under in the spring (see chapter 2).

WATER

Without moisture the soil would be practically inert. Although soils might abound in nutrients, plants couldn't soak up a single element if it weren't for the action of water, which functions as a medium through which nutrients are transferred in solution to root hairs. In addition, water helps aggregate the soil particles, acts as a solvent, and forms 85 to 90 percent of the weight of most plants. These are some of the reasons why growth seems to mushroom after a soaking rainfall.

And *soaking* is the key to good irrigation practices. Deep, penetrating watering will encourage roots to burrow deeply into the soil where moisture is more likely to be retained and available continually to plants. Shallow, timid watering, on the other hand, will tempt roots to spread out near the surface where they can fall victim to a hoe or wither from an unrelenting sun. Watering should be slow and gentle so that moisture seeps into the soil rather than running off the surface, carrying valuable topsoil with it.

Spare the Water and Spoil the Crops

Shallow waterings cause roots to stay near the surface where they are easy prey for hot sun and drought.

Thorough soakings encourage roots to grow deeply where moisture supply is more continuously available.

There are a variety of ways to get water to the garden when rainfall is scarce or where the climate is dry. The important thing is to provide water before it is too late. Don't let plants droop all the way to the ground before unreeling the garden hose. If it has not rained during the week, plan on a couple of hours of irrigation over the weekend.

Watering Can Although hardly a substitute for a gentle spring shower, the watering can will help plants through a dry spell, if your garden is no larger than about twenty feet square. Before sprinkling, scoop some soil into a saucer-like circular ridge around each plant so water sinks into the root area instead of running off. A thorough watering means several trips for each row, so plan on spending a little time.

Garden Hose One of the most useful items in the gardener's arsenal, the hose, can be a plant's best ally. Adjust the nozzle to a fine spray, then prop it up with a stone or forked stick. To prevent flooding, move the hose at least every 15 to 20 minutes. Also on the market are perforated plastic hoses that send out a mistlike spray through pinholes. When placed between rows, these perforated hoses make excellent irrigation devices.

Irrigation Ditches An effective way to provide for thorough penetration of water is by digging deep furrows between crops and filling them with water. Using boards, you can dam off sections and irrigate them separately. Or you can leave water running slowly into a central ditch that feeds into a network of other ditches. Remember to remove the nozzle and wrap the end of the hose in burlap or tie on a mitten or old sock to prevent soil from being eroded away by the force of the running water.

Sprinklers The various types of sprinklers are by far the most convenient as well as the most effective way to get water to plants. Either the rotating or oscillating type will do. Both are obtainable at hardware stores or garden supply centers.

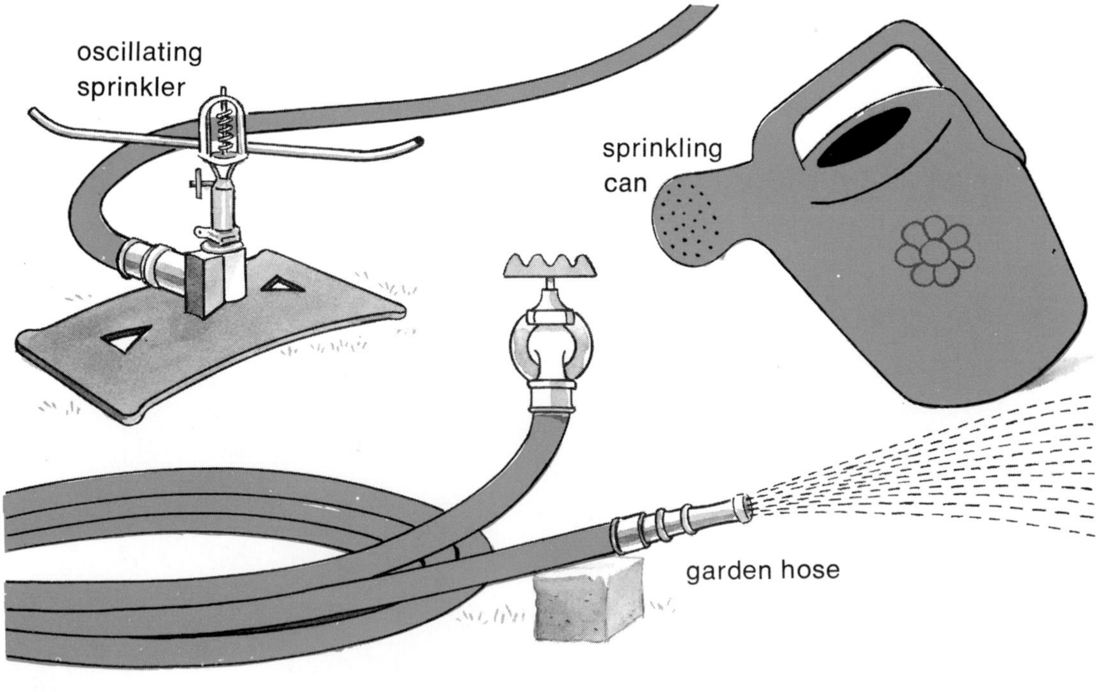

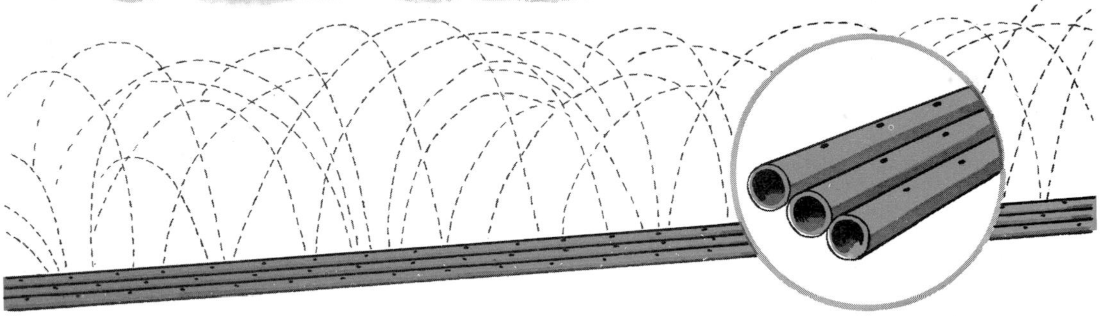

sprinkler-soaker

WHEN AND HOW MUCH TO WATER

Schedule waterings during the morning hours if possible, so plants can dry off before nightfall and lessen the chance of disease getting a foothold. Gray days are ideal for irrigation because there is less chance of evaporation. Remember too that one extended watering is better for plants than several scattered light waterings. Try to water heavily once a week—oftener in dry areas. Between one and one-and-a-half inches of water is needed to convey moisture six to eight inches below ground level where it is needed for most vegetable roots.

Plants need at least 1 inch of water (rainfall or other water) per week for good growth and abundant yields.

To keep track of how much water has been distributed over the garden, simply place several coffee cans at different locations between rows and in paths. Either place a one-inch mark inside the can with a felt pen or measure the water depth with a ruler. Since some rainfalls are deceptive, these homemade gauges will tell you exactly how much rain has fallen and whether supplemental waterings are necessary. They also measure the amounts from hose waterings. More accurate rain gauges are available at garden centers or from seed catalogs.

Unfortunately, too much water can do just as much damage as too little. A soggy soil can suffocate roots, which also need oxygen. So stick to recommended amounts when undertaking a watering program. Experience will begin to guide you as you watch results on your own grounds.

Rainfall Distribution in the United States

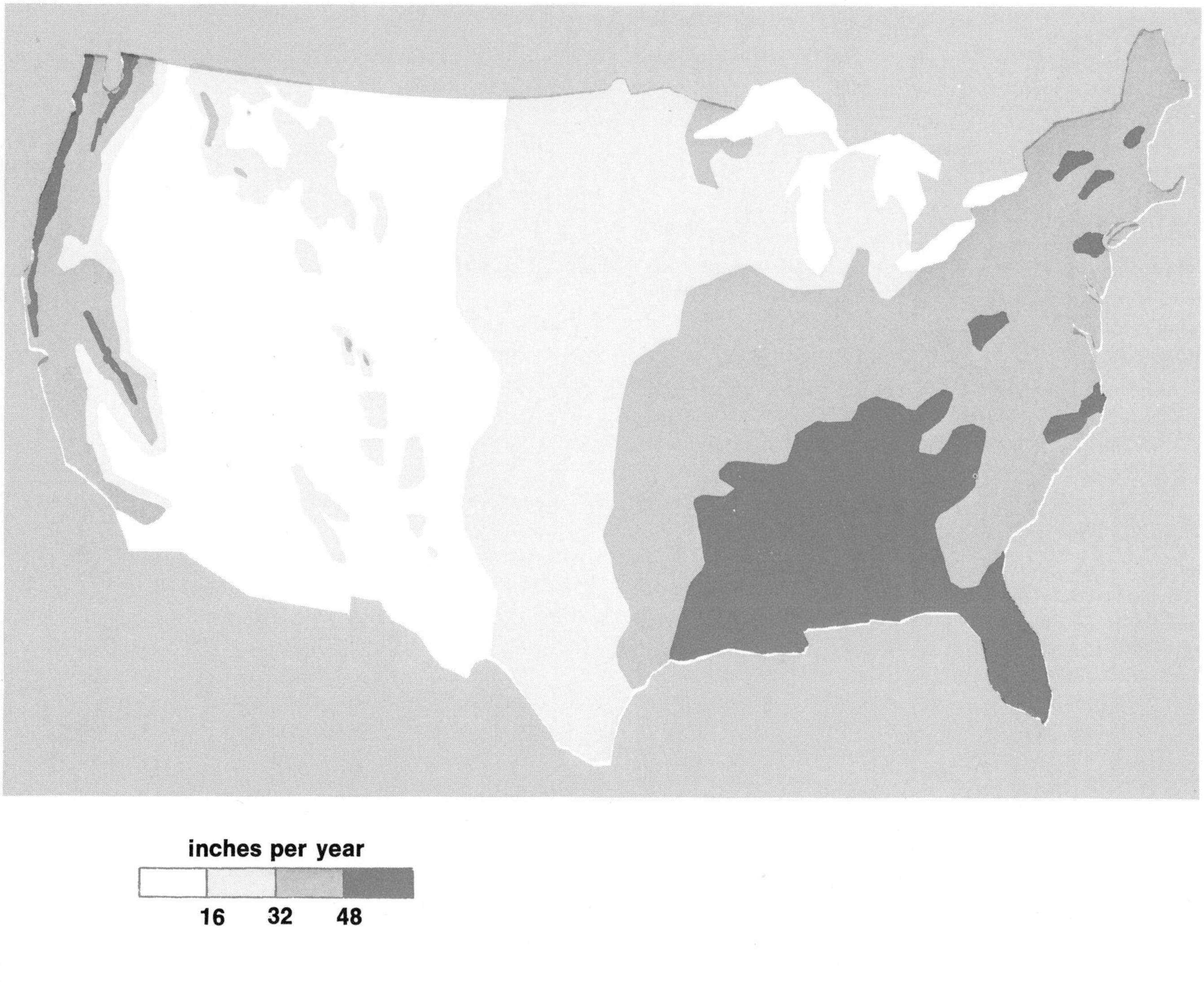

5.

Bugs and Other Battles

Bugs unfortunately have a bad press, probably because the only bugs we notice are the ones we don't want. Or we hear grisly accounts that are grossly exaggerated and stretched way beyond the facts. The truth is, bugs have a complicated system of checks and balances all their own. In the unending chain of pests and predators there are countless insects that make it their full-time business to attack and forage on other insects. Wasps, for example, solve a host of domestic problems by injecting their eggs into caterpillars or aphids, which serve not only as custom-made incubators but also as a source of food for the young when the eggs hatch. Obviously a handy arrangement for the wasp, it is also handy for gardeners because in the process the caterpillar and the aphid are killed and eliminated from the garden, while their predator lives on.

This push-pull aspect of nature, where, in a sense, one living thing keeps track of another, is what ecology is about. What the gardener must do is pause a moment and try to understand what goes on in the small sanctuaries among the leaves. He'll find that many of the menacing creatures that cause him to tuck his head between his shoulders are really friends. The dragonfly, for example, with its helicopter eyes and discomfiting habit of hovering a few inches from a man's face as if taking the dimensions of his soul, is one of his most valuable allies.

Occasionally a pest will become numerous enough to threaten a prize row of beans or corn or some other plant. If it does, don't panic, and above all resist the impulse to reach for the deadliest poison available. After all, you can always let the bugs eat their fill and then simply plant another row. The important thing is to develop an environment-conscious attitude. The chart of bugs and what to do about them at the end of this chapter is designed primarily for reference—not *all* of them will find their way into *every* garden. At worst, one or two will appear and they can be controlled simply and cheaply.

The first thing to consider, however, is prevention. A well-cared-for garden will be a healthy garden. By using the right practices, you can make your garden unappealing to harmful insects.

1. Grow the right thing in the right place. For instance, pests will be delighted if you try to grow a warm-weather crop in a cold muck soil where

the growing season is short. Because conditions are unfavorable, the plants are weaker and become perfect fodder for bugs.

2. Grow disease-resistant varieties. Remarkable strides have been made in recent years in developing strains resistant to disease. A good seed catalog or your State Agricultural Experiment Station is a good source of information concerning resistant varieties.

3. Keep weeds on the run. A thicket of weeds can provide an ideal haven for pests as well as a jumping-off point for diseases. Keep the area immediately surrounding the garden free of weeds by periodic cutting or mowing.

4. Inspect store-bought plants carefully. Before giving up cash for plants, examine them for root swellings (on plants where roots are visible), cankers, and spotty leaves. A healthy plant stands tall and firm and has a lively green color to its leaves.

5. Stay out of the garden when plants are wet. Give them a chance to dry off before going about your gardening chores. Certain waterborne diseases can infect the entire garden by hitchhiking on your hands or clothing.

6. Destroy plant residue after harvesting each crop. As soon as possible after the harvest is in, carry the plant remains off to the compost heap, where high temperatures will stamp out spores or larvae that might otherwise winter over and emerge the following season. Or, if the residue is not infected, use a rotary tiller to shred the waste material right into the garden soil.

7. Rotate disease-prone crops yearly to different locations. Cabbage, Brussels sprouts, cauliflower, and broccoli, for example, are susceptible to certain root diseases and should be given a different location each season.

8. Encourage beneficial birds and insects. Birds, toads, ladybugs, praying mantises, and a host of other coinhabiters of the earth are potential allies of the gardener and should be offered the welcome mat when they appear.

9. Use only disease-treated seeds. Those from reliable seedsmen will usually be treated to prevent certain soil-borne diseases from affecting them. Most seed packages indicate whether the contents have been treated.

10. Remove and destroy diseased plants immediately.

11. Mulch plants to prevent disease spores lurking in the soil from splashing up on the plants in rainstorms.

KNOW YOUR FRIENDS

Scattered through the crowd of bugs buzzing through a June afternoon are a number of positively helpful insects that are of great value to the home gardener. Before you take spray in hand, become as familiar as you can with the good guys.

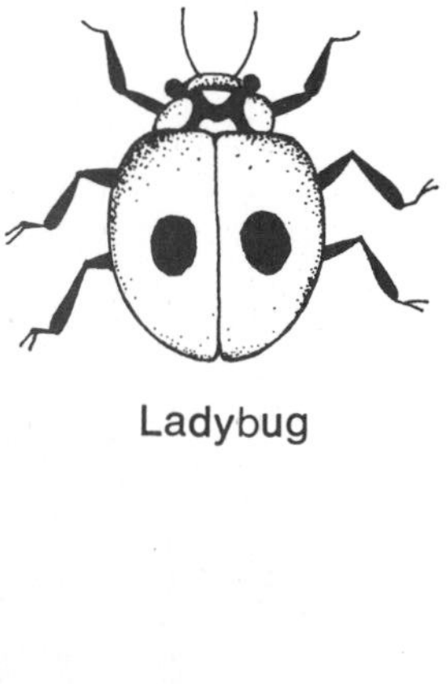

Ladybug

Ladybug (also called Ladybird and Lady Beetle) The spotted red backs of lady beetles are a familiar sight to everyone. But not everyone appreciates their eating habits. A single ladybug can easily go through as many as 50 aphids in a single day's foraging. Ladybug larvae, although not much to look at, are also enemies of the injurious aphid. If the ladybug population in your area appears low, you can order a supply from a local garden supply center or by mail from a good seed catalog, and distribute them in and around the garden area.

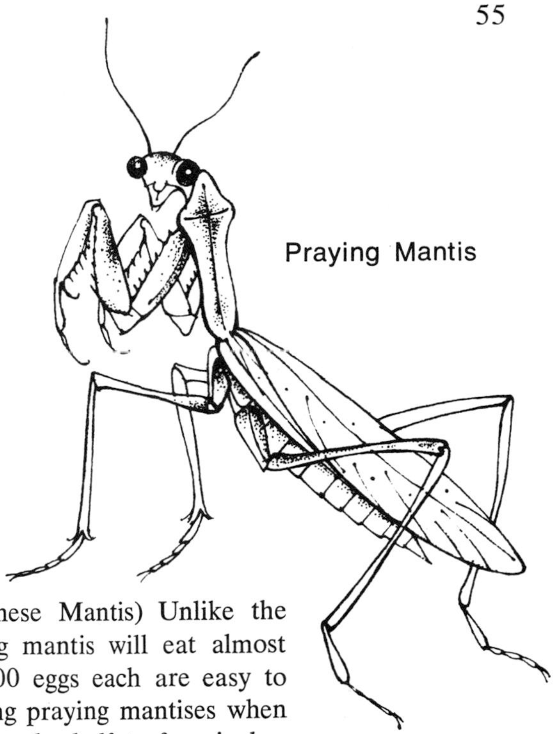

Praying Mantis

Praying Mantis (also called Carolina Mantis and Chinese Mantis) Unlike the ladybug, whose diet is limited to the aphid, the praying mantis will eat almost any insect that moves. Egg masses containing 100 to 300 eggs each are easy to obtain from a seedsman and will produce crowds of young praying mantises when attached to a shrub or bush. They move slowly, are two-and-a-half to four inches long, and are a light green (some are brownish) in color. Egg masses are the size of a small hen's egg, lumpy, brown, attached to a twig or stem. *Do not destroy them.* They will hatch in spring.

Beetles Of the thousands of known beetle species, it is the ground beetle that is most beneficial to garden crops. Dark brown or black in color, with a shield-shaped body, it hides in the soil during the day and emerges at night to rummage for insects; it also feeds on snails and cankerworms, larvae and caterpillars. Other beneficial beetles include the European ground beetle, unmistakable because of its iridescent blue-green color, and the tiger beetle, also iridescent but of a distinct blue, green, or bronze hue.

Robber Fly The robber fly appears wingless and is somewhat startling to the gardener because of its quickness. The soil-borne larvae feed on other larvae.

Assassin Bug Similar to the praying mantis in its method of catching prey, the assassin bug crawls lethargically along plant leaves and stems in search of food. Usually brown in color with long wings and less than an inch long, it helps to control aphids, leafhoppers, and caterpillars.

Robberfly

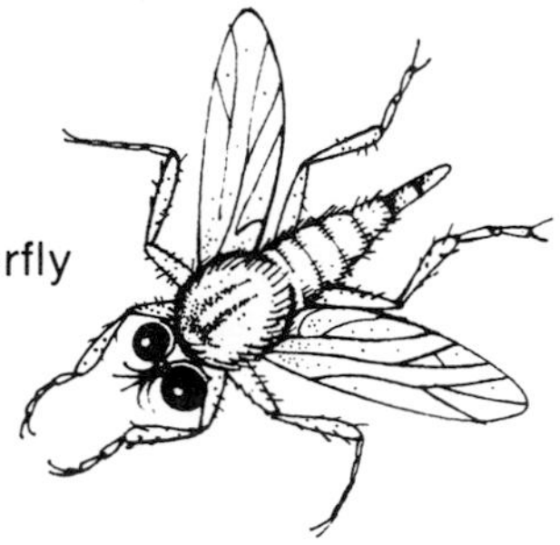

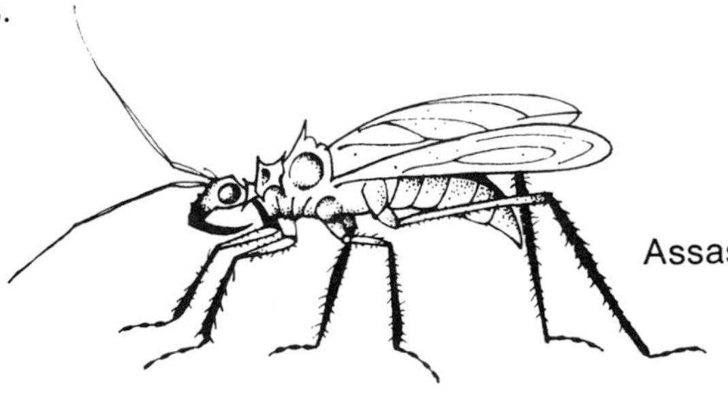

Assassin Bug

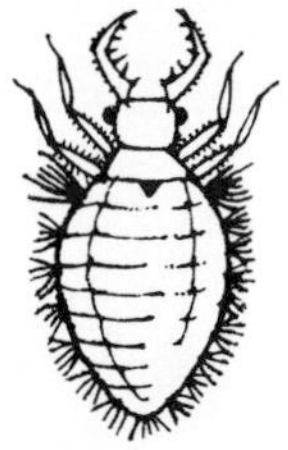

Ant-Lion

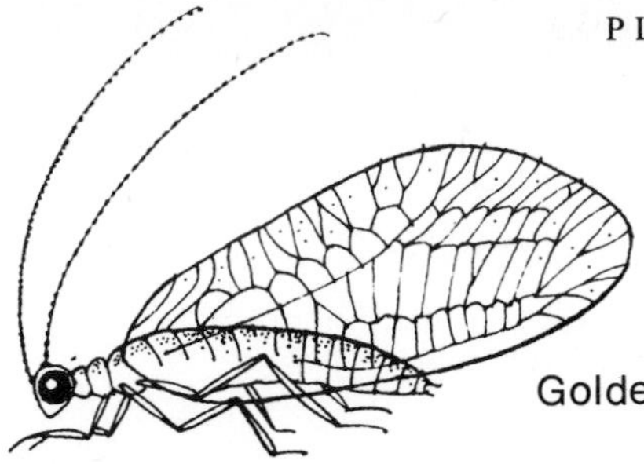

Golden-Eyed Lacewing

Golden-Eyed Lacewing (also called Aphid Lion) The golden-eyed lacewing is a common visitor to the garden and is easily identified by the golden hue of its eyes and its translucent gauzy green wings. Because the homely looking larvae are especially active seekers of aphids, they have earned their other name of aphid lion, often consuming 200 to 400 aphids before reaching maturity.

Ant-Lion (also called Doodlebug) The larva of the ant-lion carefully digs a cone-shaped pit with smooth sides, then waits at the bottom for its prey—ants and other victims—to tumble in. The adult ant-lion closely resembles a dragonfly.

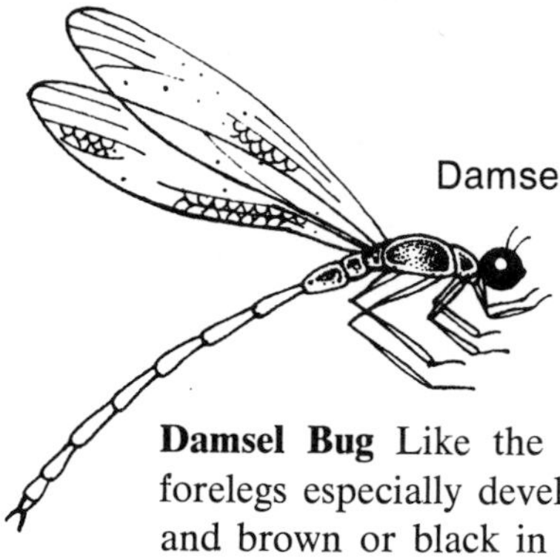

Damsel Bug

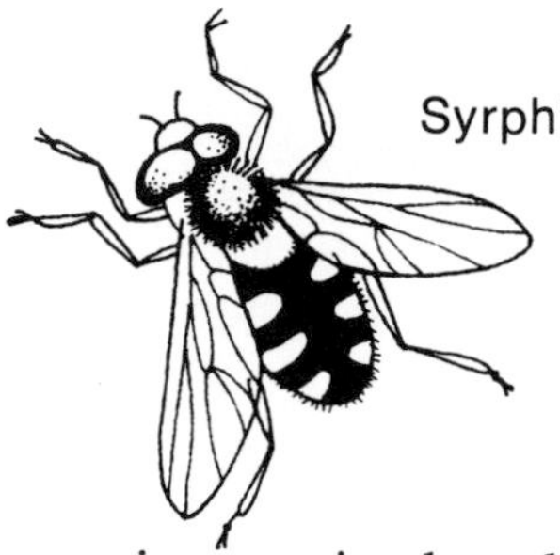

Syrphid Fly

Damsel Bug Like the assassin bug and the praying mantis, damsel bugs have forelegs especially developed for catching prey. About three-eighths inch in length and brown or black in color, damsel bugs feed on aphids, fleahoppers, and small larvae of insects.

Syrphid Fly (also called Flower Fly) Syrphid flies are hovering insects whose larvae are unquenchable aphid hunters. A single larva at full tilt is said to consume an aphid a minute. The larvae are sluglike in appearance and may be brown, gray, or mottled.

Braconid Wasp

Wasps Although dreaded because of its sting, the ordinary wasp is of great value when it comes to controlling garden pests. The female Braconid wasp, for example, injects her eggs into living aphids, the eggs of host insects or parasites of caterpillars, and larvae of other insects. When the wasp larvae emerge, they begin feeding on fatty tissue and eventually kill the host. Other wasp species such as yellow jackets and paper wasps feed directly on scale insects and their eggs as well as on grubs and caterpillars.

This list of beneficial insects whose habits coincide with the interests of the gardener is hardly complete. Spiders with their webs and certain mites can also contribute to pest control. It is the tug of war between pest and predator that prevents one insect species from taking over the earth. Before buying a single bug bomb, spray, or dust, learn to respect and appreciate the forces in nature that are on your side. You'll be a better gardener because you will use the lethal chemicals with more restraint as you work in—and with—nature.

SOME COMMON GARDEN PESTS

Acting in haste and out of ignorance and wiping out a beneficial insect is a disaster that will be difficult for your garden to overcome. Try to know the true enemy. Most harmful insects fall into two broad categories, depending on their feeding habits: chewing insects and sucking insects.

Generally larger than sucking insects, the chewing insects attack leaves, especially the young tender leaves near the tip of the plant. Pinholes, irregular holes throughout the leaf, or chunks bitten from the edges are clear signs that a chewing insect has been at work. The Colorado potato beetle, the cabbage looper, and the flea beetle are examples of leaf chewers that can easily defoliate a plant if left unchecked.

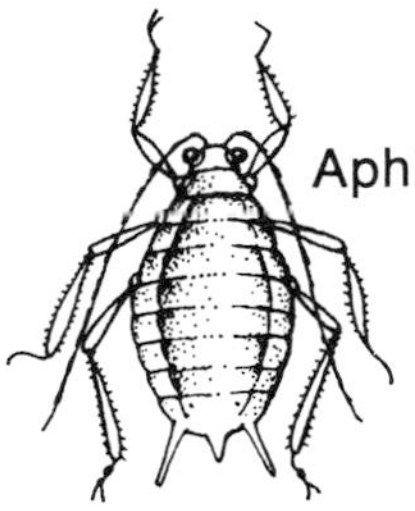

Aphid

Sucking insects have needlelike mouth parts that pierce the leaf and suck out juices. They are not always easy to detect, since they often cling to the undersides of leaves to escape predators. But evidence of their presence is obvious —the leaves begin to curl and shrink at the edges and eventually die. The small droplike aphid is by far the most common sucking insect to attack plants.

Aphids The most numerous of garden pests, aphids cling in clusters to the undersides of leaves or to succulent shoots, and silently suck juices from the plant. Because of their small size and light green to powdery blue color, which blends into foliage, they are often invisible. Wasps and ladybugs are the most efficient biological controls for aphids. In many cases a strong blast of water from the garden hose can scatter a crowd of aphids so widely that they won't return. If all else fails, a rotenone, pyrethrum, or nicotine spray will do in aphids without wiping out ladybugs.

Flea Beetle

Flea Beetles The small black flea beetle has powerful legs that catapult it to safety when it is disturbed. Affected leaves are easy to spot because of the abundance of pinpoint-sized holes that look as if they might have been caused by a blast of tiny buckshot. About one-sixteenth inch long and sometimes striped, flea beetles are especially fond of tomato, potato, and eggplant leaves, but will eat whatever foliage is available. Sevin or endosulfan will discourage flea beetles.

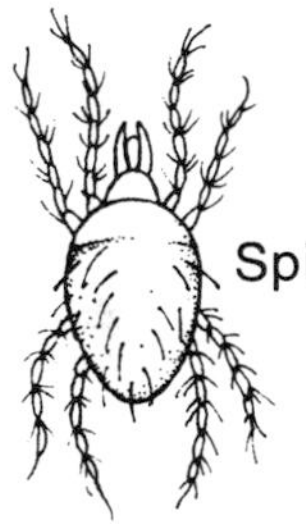

Spider Mite

Cutworms Cutworms can be the most heartbreaking of all the garden pests. In hiding during the day, they emerge at night to slice tender seedlings off at ground level as if with a razor blade. Dirty brown or gray in color, they are easily foiled without using a poison by wrapping paper or aluminum-foil collars around newly-set-out plants.

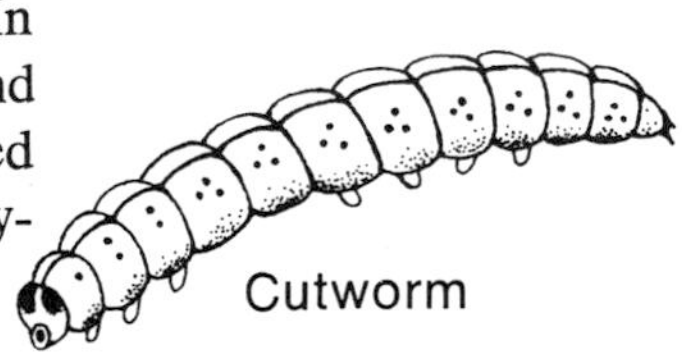

Cutworm

Spider Mites These mites are almost invisible to the naked eye, but the telltale sign of a spider mite infestation is numerous yellow specks on leaves and stunted plants. Usually natural predators will keep spider mites in check. If not, try dicofol or a spray containing sulfur.

Leaf Miners The leaf miner forages within the leaf itself, its wirelike tunnels causing blotches and blisters on the leaves. Attracted to pepper and spinach leaves, the fly larvae are yellow in color and about one-eighth inch long. A spray containing Diazinon or dimethoate can offer some control.

The controls listed here are by no means complete. Check the chart at the end of this chapter for effective controls for insects that attack specific crops.

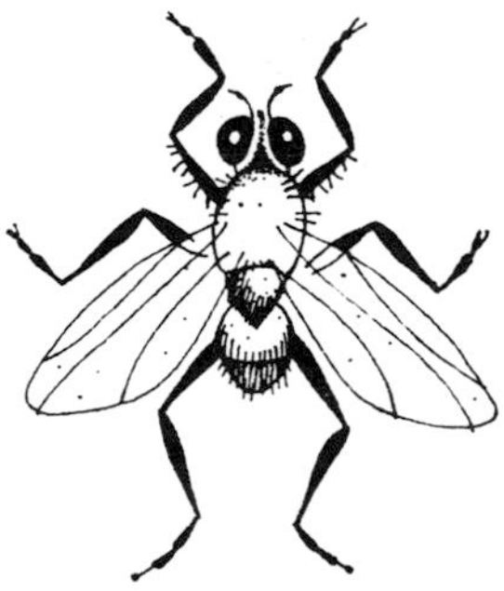

Leaf Miner

WHAT TO DO

If you come across a bug in the garden, resist the initial waves of panic and ask yourself a few questions before reaching for a poison spray.

Is the bug just a loner or does it have a crowd of friends?

Is it actually doing some damage or just passing through?

Is it chewing on leaves or on the pesty aphid?

Does it match the description of any of the beneficial insects?

Once you have matched the insect with the injury and are fairly certain that the crop is threatened, the next step is to decide on a control measure. But force yourself to think in terms of temporary control and not of blasting the pest off the face of the earth forever. Play it safe by first trying a biological control, which uses no poisons whatsoever. If that is impractical or ineffective, then turn to those approved chemical controls that are clearly noninjurious to the environment. Only as a last resort should you pick a toxic, highly poisonous pesticide, and then spot-blast *carefully*.

Biological Controls Nature's biggest weapon in insect control is climate. A long hard winter with subzero temperatures can practically eliminate some species that have overpopulated a particular area. Of course, we have little control over the weather and you may inhabit an area that is not cold. But you can search out those insects that include other insects in their diets (see page 54). In addition, it is known that some plants seem to be naturally repellent to certain insects, and these can be used to advantage. While scientific proof does not yet exist—this concept is still new and needs exploration—many plants are clearly unattractive to certain insects. Try some of these companion plantings before opting in favor of a poison:

BEANS planted near or among potatoes are fairly effective in discouraging the Colorado potato beetle. At the same time the potatoes will act as a repellent to the Mexican bean beetle.

CHIVES AND GARLIC should not be isolated but scattered among lettuce or pea plants as a deterrent to aphids.

TOMATOES planted adjacent to asparagus can effectively contain the asparagus beetle population.

RADISHES sown in hills of cucumbers are abhorrent to the striped or spotted cucumber beetle.

TANSY planted in and around cabbage plants will put a stop to the ever-present cabbage worm and discourage the destructive habits of the cutworm.

GERANIUMS can be cultivated near grapes to thwart Japanese beetles.

MARIGOLDS scattered among potato plants may be enough to discourage the Colorado potato beetle.

Another way to clamp down on harmful bugs is to become a predator yourself and hand-pick pests off plants. Beetles and other hard-shelled insects can be killed by immersing them in kerosene, while smaller, softer bugs can be simply dropped into a container of hot water. For the European corn borer, hand-picking is often the most effective method of control available. Look for telltale sawdust piles at the joints of leaves for signs of this troublesome insect. Broken corn tassels also indicate their whereabouts. Using tweezers or the fingers, squeeze as many borers as you can find.

Chemical Controls Many chemicals have become known as villains responsible for doing serious damage to the environment. But it is worth remembering that chemicals cannot spread themselves over crops—it is man's use of chemicals that does the damage. All too often we have coated the earth with poison as if in a blind rage, never considering the aftereffects. Give nature a break by taking the trouble to understand *what* sprays contain, *how* to use them, and *when* to use them. Very rarely will pests become so threatening that a powerful chemical is necessary. Also, the various chemicals can be grouped according to how harmless or harmful they are and on what pests or diseases they are effective. Begin with the least toxic, and if that fails gradually work your way up the scale.

HARMLESS MATERIALS

Water Believe it or not, water can often provide all the control you need. As already indicated, a strong blast of water aimed directly at plant-sucking aphids will knock them off. Chances are they would rather move on than return for more. Stubborn pests can be killed outright by dousing plants with warm water.

Oil Available in stock preparations from garden centers, oil is easy to use and effective against undesirable insects. Certain dormant sprays for fruit are based in emulsion. Follow the directions provided for best results.

Salt Solutions These offer a handy, completely safe method of combating spider mites and cabbage worms. Dissolve one ounce of salt in a gallon of water and spray plants with a standard sprayer designed for vegetable crops. Take care not to saturate the soil with the mixture, since salt can rob moisture from plant roots and too much will kill tender plants. *Don't* increase the dose.

RELATIVELY SAFE MATERIALS

(See chart on page 64 for specific uses)

Rotenone One of the most reliable natural materials for pest control, rotenone is produced from grinding the roots of the derris plant. Sometimes also sold as Derris, it is an effective contact insecticide and relatively safe to man and animals (except fish, for which it is highly toxic). Rotenone will discourage and control most chewing and sucking insects as it has some residual effect by remaining on leaves for a short time.

Pyrethrum An extract from plants of the chrysanthemum family, pyrethrum is an old stand-by. Effective as a contact insecticide for the instant killing of a host of vegetable and fruit pests, pyrethrum, together with rotenone, should be all you need. It too is safe to man and animals, except fish, and does little harm to bees and beneficial insects.

Bacillus Thuringiensis (Bio-guard, Thuricide) This is especially useful for combating tomato hornworm, canker worms, and cabbage worms. Essentially it's a microbe disease that works by infecting the pest. As a microbial insecticide it acts slower than a chemical insecticide but is just as effective. It's available in either liquid or powder form.

Lime-Sulfur Primarily used as a dormant spray to combat scale insects and mites, lime-sulfur is recommended mostly for the control of scab on fruit trees.

Bordeaux Mixture This copper fungicide, developed in France to protect vineyards, is a safe and effective antidote for many diseases and a few insects. Available as a wettable powder, the mixture is dissolved in water and sprayed on plants as a fungicide. It is especially useful against downy mildew.

You might have difficulty obtaining some of these materials. Because we have grown used to demanding quick-and-easy solutions to problems, the market is glutted with poisons that certainly kill pests but, unlike the controls presented here, kill a lot of other things as well. Keep asking for these sprays, even if you cannot get them. Sooner or later, manufacturers might just heed the public's demand for safer, environment-protecting substances. Meantime these are the generally approved controls.

TO BE USED WITH CAUTION

Carbaryl (Sevin) Considered an acceptable stand-in for DDT, which is outlawed in most places, carbaryl checks chewing and sucking bugs such as leaf miners, lace bugs (not to be confused with the beneficial lacewing), and chinch bugs. It is, however, residual and should not be applied to fruits and vegetables less than three days before harvest. Sevin is toxic to bees and should be replaced by methoxychlor if pollination is critical.

Malathion Also a general insecticide, malathion should not be applied within a week of harvest date.

Methoxychlor Especially effective against chewing insects, methoxychlor will also accumulate in the human body so care must be taken in applying it. It should be used no later than two weeks before crops are harvested.

Kelthane Kelthane is a useful miticide, if it is mites that are causing you problems. At least a full week should intervene between the last application and harvesting of crops.

As concern for living things—including humans—mounts, the list of poison materials approved by federal and state agencies is constantly changing. At the same time, manufacturers are developing newer and safer substances. Write your State Agricultural Experiment Station for the latest listing of approved sprays.

ALL-PURPOSE SPRAYS

Since there are at least as many pests as there are fruits and vegetables, you may be faced with a variety of bugs cropping up at different times during the growing season. An all-purpose spray, consisting of two or three insecticides mixed together, can often be the wisest solution to the bug and disease problem. A number of combination sprays are on the market, but if you have trouble finding one, try this versatile mixture that you can put together yourself:

1 tablespoon	**Rotenone** (emulsifiable concentrate)
2 tablespoons	**Captan** (wettable powder)
3 tablespoons	**Zineb**
1 gallon	**Water**

OR

2 tablespoons	**Malathion**
2 tablespoons	**Methoxychlor**
2 tablespoons	**Zineb**
1 gallon	**Water**

Put the powder into a small sprayer and mix with a small amount of water to form a paste. When the powder is dissolved add remaining water. Stir often during use to prevent settling. Keep leftover spray in a clearly labeled, tightly-capped glass container, and clean the equipment thoroughly before storing.

Keep *all* insecticide powders and liquids in their original containers and lock them up out of the reach of children.

DISEASES AND OTHER PROBLEMS

Occasionally a plant will collapse into a pathetic slump without the presence of a single bug. Chances are it is under attack from bacteria, fungus, or virus. Fortunately there are a few effective control measures. The first is to follow the preventive gardening practices listed at the beginning of this chapter. Many fungi can be nipped before too much damage is done with a commercially prepared fungicide applied as a dust or spray.

Damping Off Damping off is caused by a fungus in the soil that can destroy almost any young seedlings by attacking at ground level with a kind of rot. Unchecked, a siege of damping off can flatten a stand of seedlings in no time. Most commercially available seed has already been treated against damping off.

If not, follow these seed-treating directions:

1. Tear off one corner of the seed packet.
2. With a small knife blade, drop a small amount of fungicide dust (captan) into the packet.
3. Fold down the corner of the packet.
4. Shake vigorously to distribute the dust over all seeds.

If you suspect the presence of fungi in your outdoor soil, water the row with captan just after planting.

When planting seeds indoors, be sure to use a sterilized soil. Either bake ordinary garden soil in the oven (200 degrees for 45 minutes) or use a compost-soil mixture topped with a half-inch layer of vermiculite.

Early Blight Small, irregular dark-brown spots on leaves that grow into large spots with targetlike rings indicate the presence of early blight. Stems may develop brown cankers at the soil line, while dark, leathery, decayed spots may appear at the stem end of fruits. This disease is prevalent in moist warm regions. When buying plants be sure none of these symptoms are present. If the blight develops, spray with zineb, captan, or another fungicide every seven to ten days.

Leaf Spot The disease appears first as small spots with light centers and dark edges. The leaves eventually die and yield is drastically reduced. Warm and moist conditions encourage the disease. As a preventive measure, rotate crops from year to year and destroy plant residue after each harvest. Affected plants may be sprayed with any good fungicide.

Late Blight The first sign of late blight is dark water-soaked spots on leaves and fruit. Occasionally a white powdery growth will appear on the undersides of leaves and sometimes on fruits as well. This fungus is fond of cool moist weather. Spray immediately with fungicide, repeating treatment every seven to ten days.

Anthracnose Numerous reddish-brown circular spots on leaves, elongated cankers on stems, and sunken spots on fruits make anthracnose easy to detect. A common disease of cucumber, watermelon, and muskmelon, anthracnose fungus can be troublesome in warm moist areas. To control it, allow at least two years to elapse before growing cucumbers or melons in the same soil. Use a fungicide spray or dust, being sure to follow the directions on the label carefully.

Downy Mildew Yellow angular spots appearing on older leaves are a sign of downy mildew. Eventually the leaves will dry out, curl, and die. Use mildew-resistant varieties of cucumbers and melons, the plants most commonly affected. A dust or spray will control mildew on existing plants.

Rust This fungus commonly attacks asparagus and can seriously reduce the following year's crop if not held in check. Symptoms are wormlike orange pustules on stems and foliage. Since the fungus can winter over in the remains of diseased plants, it is wise to destroy all residue after harvest. Study seed catalogs for rust-resistant varieties but, should rust appear, make sure the diseased plants are cut at ground level and burned in the fall.

For pests and diseases affecting specific crops, their symptoms, and what to do, see the chart at the end of this chapter.

PRECAUTIONS–A SUMMING UP

The various insecticides have one purpose—to kill, by poisoning either through ingestion (chewing insects) or through contact (sucking insects). Remember that a spray or powder can't tell the difference between good and bad bugs, or animals, pets, and children—it will affect whatever it comes in contact with. A cardinal rule is: **Always Read the Label and Follow Directions Carefully.** After all, the manufacturer who made the insecticide is obviously in the best position to tell you how to use it and what NOT to do. Study mixing and application directions thoroughly before using it.

Similarly, fungicides and other controls should be handled with respect and all due caution. It cannot be said too often that they should *all* be kept locked up where children, pets, and even unauthorized adults cannot reach them.

> Be sure you can identify the culprit. If, after consulting the list of common pests and the disease and pest chart, the bug still remains a mystery, show it to your State Agricultural Experiment Station or county agent. Not until the bug has been identified can you make an intelligent, environment-protecting choice of pesticide.
>
> Make sure the poison fits the crime in order to avoid overkill. By all means try the least potent controls first (water, oil, a weak salt solution, biological controls, companion plantings) before reaching for stronger, more persistent insecticides.
>
> Keep in mind that ultimately you may eat what you spray. Some sprays and dust have no residual effect and can be applied within a day of harvest. Others may be frighteningly residual and will collect in the fatty tissues of the body if ingested. Follow package directions exactly.
>
> Store materials and equipment in a safe place. Clean the sprayer thoroughly after each use and keep poisons in their *original* containers. Do not dump leftover spray where it can wash away and collect in pools from which birds and animals may drink.
>
> Avoid spraying or dusting on windy days. Also, watch out even on still days for the drift of sprays from what you are treating onto nearby food crops you are about to harvest. If you have prized flowers or shrubs that require constant spraying, plant your vegetables where they will not be in the way of the drifting sprays.

The following chart covers in short form all the major pests and diseases that afflict vegetable and fruit crops, and is a handy reference aid to identification as well as treatment.

GARDEN PESTS AND DISEASES AND CONTROL MEASURES

CROPS	DISEASE/PEST	SYMPTOMS	CONTROLS: NONHAZARDOUS	CONTROLS: SOME HAZARDS	APPLICATION
MOST CROPS	*Aphids* (small tear-shaped insects clustering on undersides of leaves and on ends of new shoots)	Curled leaves, discoloration, and wilting	Water from garden hose or combined with a *nondetergent* soap or salt (1 oz. to 1 gal. water)		Blast off insects with stream from garden hose or saturate plant with mixture.
			OR 5% Rotenone WP (5T to 1 gal. water) or 1% Rotenone dust	25% Diazinon EC (2*t* to 1 gal. water) 57% Malathion EC (2*t* to 1 gal. water)	Spray plants thoroughly including undersides of leaves. Check container label to see if plants are listed in manufacturer's instructions.
	Beetles: *Blister Beetles* (gray, black, or striped, ½ to ¾ inch long) AND *Flea Beetles* (pinhead-sized, black, brown, or striped, jumping insects, 1⁄16 inch long)	Chewed leaves; Numerous buckshot-like holes in leaves, which weaken plant and invite disease	5% Rotenone WP (5T to 1 gal. water)	50% Sevin WP (2T to 1 gal. water)	Cover plants as thoroughly as possible when beetles appear. One or 2 applications are usually sufficient.
	Cutworms (dull gray, brown, or black, sometimes striped or spotted; up to 1¼ inches long; curl up when disturbed)	Plants cut off cleanly at soil level as if with razor blade	Place 3-inch cardboard, aluminum foil, or stiff paper collar around each plant and insert 1 inch into soil	10% Toxaphene dust OR 60% Toxaphene spray (1T to 1 gal. water)	Apply insecticide to the soil surface when garden is being prepared for planting.
	Damping-Off Disease (soil-borne fungus that attacks stems of seedlings at soil level)	Pinched, discolored stems of seedlings at ground level; plants flatten out and die	When planting seed in flats use sterile soil; use reliable or treated seed and avoid excessive moisture	Treat seed with fungicide if not already treated (see p. 62) 50% Captan WP (1T to 1 gal. water)	Water seedlings once per week until 10 days before transplanting.

			NONHAZARDOUS	SOME HAZARDS	
	Leaf Miners (yellow larvae ⅛ inch long living within leaves)	Slender white tunnels inside leaves		25% Diazinon EC (2*t* to 1 gal. water) OR 57% Malathion EC (2*t* to 1 gal. water)	Spray on leaves as soon as small tunnels are noticed. Allow at least 7 days between last spraying and harvest.
	Spider Mites (barely visible red or greenish-red, found on undersides of leaves)	Fine web structures and yellow spots on leaves	30% sulfur dust or spray containing sulfur	35% Kelthane WP (1T to 1 gal. water) OR 18.5% Kelthane EC (2*t* to 1 gal. water)	Spray both sides of leaves when mites can be seen or when general off color is noted. **Caution:** Sulfur can cause damage to cucumber and melon plants.
Asparagus	*Cutworms* (see listing under MOST CROPS)	Pinched or crooked spears		50% Methoxychlor WP (6T to 1 gal. water) OR 50% Sevin WP (4T to 1 gal. water) OR 5–10% Sevin dust	Dust or drench soil before seedlings emerge or when plants are set. In established beds apply just around plants when damage is noticed.
Beans	*Mexican Bean Beetles* (oval and copper-colored, ¼ inch long, with 16 black spots from left to right on wing segments)	Damaged pods and leaves, eventually skeletonized	Hand-pick in small plantings OR 5% Rotenone WP (5T to 1 gal. water)	50% Sevin WP (1T to 1 gal. water) OR 57% Malathion EC (2*t* to 1 gal. water)	Spray when beetles appear, being sure to cover undersides of leaves.
	Bean Leaf Beetles (red to yellow with black spots, up to ¼ inch long)	Feeds on leaves, causing regular-shaped holes in foliage	Hand-pick pest OR 5% Rotenone WP (5T to 1 gal. water)	50% Sevin WP (2T to 1 gal. water) OR 57% Malathion EC (2*t* to 1 gal. water)	When beetles appear spray entire plant thoroughly, including undersides of leaves.

EC Emulsifiable Concentrate (liquid)
WP Wettable Powder
T Tablespoon
t teaspoon

CROPS	DISEASE/PEST	SYMPTOMS	CONTROLS		APPLICATION
			NONHAZARDOUS	SOME HAZARDS	
Beans (*Cont'd.*)	*Seed Maggot* (yellowish-white worm, ¼ to ⅓ inch long)	Bores into seeds, damaging or destroying emerging plants	Use reliable treated seed; plant when soil is thoroughly warmed in shallow furrows; keep manure applications light; remove infested plantings and sow again	50% Diazinon WP (1T to 1 gal. water)	Drench seed furrows or hills before planting.
	Anthracnose, Downy Mildew, and *Rust* (all fungi)	Leaf discoloration, sunken spots, or pustules (blisters)	Rotate crops; 30% sulfur dust or spray containing sulfur; use resistant varieties	80% Maneb WP (2T to 1 gal. water) OR 75% Zineb WP (2T to 1 gal. water)	Apply at 4- to 10-day intervals for as long as necessary. Allow 7 days between last spraying and harvest.
Beets and Chard	*Leaf Spot* (fungus)	Leaves covered with small round spots with light-colored centers	Use treated seed and rotate crops	75% Zineb WP (2T to 1 gal. water)	Spray plants when symptoms appear. One application is usually sufficient.
	Leaf Miner (see listing under MOST CROPS)				
Broccoli, Brussels Sprouts, Cabbage, Cauliflower, and Chinese Cabbage	*Aphids* (see listing under MOST CROPS)				
	Cabbage Worm or Looper (pale green, up to 1½ inches long, loops or doubles up when crawling) AND *Webworm* (dull grayish-yellow with purplish stripes, up to ½ inch long)	Feeds on undersides of leaves, producing large ragged holes. / Bores into buds and stems, killing young plants	Hand-pick loopers and immerse in kerosene or hot water OR Apply biological insecticide *bacillus thuringiensis* according to manufacturer's directions	57% Malathion EC (2*t* to 1 gal. water)	Apply before heads have begun to form and cover undersides of leaves thoroughly.

			NONHAZARDOUS	SOME HAZARDS	
	Cabbage Maggot (yellowish-white, ¼ to ⅓ inch long)	Tunnels into roots and stems of young plants; plants wither and die	Rotate crops; avoid using untreated seed	2% Diazinon granules (6 oz. per 100 sq. ft.) OR 50% Diazinon WP (1T to 1 gal. water)	Before planting seed, mix granules thoroughly with upper 4 inches of soil. When setting out transplants water each with 1 cup of spray mixture.
	Clubroot (slime mold inhabiting soil)	Large swellings, growth, or "clubs" on roots; stunted plants	Sow seed in clean soil; rotate crops; avoid areas that have produced clubroot		
Brussels Sprouts (*See Broccoli*)					
Cabbage (*See Broccoli*)					
Carrots, Parsnips, and Turnips	*Carrot Caterpillar* (green with black and yellow markings, up to 2 inches long)	Chews leaves, destroying tops	Hand-pick caterpillars when they appear	Seldom numerous enough to warrant spray or dust	
	Rust Fly (larvae are yellowish white and up to ⅓ inch long)	Bores into roots destroying crop		2% Diazinon granules (6 oz. per 100 sq. ft.)	Before sowing seed work granules into upper 4 inches of soil. Wait 1 week before planting.
	Wireworm (yellow to white with dark head and tail, ½ to 1½ inches long)	Tunnels into stems, roots, causing severe damage if not crop failure	Plant winter rye and turn under in spring (soil bacteria attacking rye will also attack wireworm)	2% Diazinon granules (6 oz. per 100 sq. ft.)	At least 1 week before planting seed work granules into upper 4 inches of soil.
	Leaf Blight (fungus)	Black or brown spots on leaves and stalks	Rotate crops; avoid excessive moisture	80% Maneb WP (2T to 1 gal. water)	Spray when plants are 6 weeks old and continue at 7- to 10-day intervals.

EC Emulsifiable Concentrate (liquid)
WP Wettable Powder
T Tablespoon
t teaspoon

CROP	DISEASE/PEST	SYMPTOMS	CONTROLS: NONHAZARDOUS	CONTROLS: SOME HAZARDS	APPLICATION
Cauliflower, (*See Broccoli*)					
Celery	*Celery Leaf Tier* (green, up to ¾ inch long)	Chews holes in leaves and stalks; encases leaves in webs	Pyrethrum dust containing 0.2% pyrethrins		Make 2 applications spaced ½ hour apart, the first to dislodge the tier, the second to kill it.
	Aphids (see listing under MOST CROPS)				
	Early Blight (fungus)	Small yellowish-brown spots on old leaves, gradually becoming larger	Rotate crops; avoid excessive moisture; destroy plant residue; use disease-resistant variety (Emerson Pascal)	75% Zineb WP (2T to 1 gal. water)	Spray affected plants. **Caution:** Trim and wash plants thoroughly before using to remove possible spray residue.
	Late Blight (fungus)	Yellow spots on older leaves and stalks			
Chard (*See Beets*)					
Chinese Cabbage (*See Broccoli*)					
Corn	*Corn Earworm* (green, brown, or pink with light stripes along sides and back, up to 1¾ inches long)	Chewed silks and damaged kernels at tip of ear	With eyedropper place 2 or 3 drops of mineral oil into each ear when silks appear	50% Sevin WP (2T to 1 gal. water) Dust not recommended	Apply the day after silks appear and repeat 4 times at 2-day intervals.
	European Corn Borer (whitish, pink, or brown with dark brown head, up to 1 inch)	Broken tassels, single hole bored into side or base of ear, or small dustlike pile of shavings at leaf axils	Hand-pick borers; check stalks each day for signs of insects	50% Sevin WP (2T to 1 gal. water)	Apply to ear shoots and into leaf whorls. Repeat for at least 3 applications at 5- to 7-day intervals.

			NONHAZARDOUS	SOME HAZARDS	
Cucumbers, Muskmelon, Pumpkin, Squash, and Watermelon	*Striped Cucumber Beetle* (yellow to black with 3 black stripes down back, ⅕ inch long; larva whitish with brown head and tail, ⅓ inch long)	Beetles feed on leaves, stems, and fruit and spread bacterial wilt; larvae tunnel into stems and roots below soil	Hand-pick beetles as they appear OR 5% Rotenone WP (5T to 1 gal. water) OR 1% Rotenone dust	57% Malathion EC (2*t* to 1 gal. water) OR 50% Sevin WP (2T to 1 gal. water)	Begin spraying as plants emerge and continue weekly, especially after rains. Apply during early morning hours to avoid injury to bees and blossom set.
	Pickleworm and *Melonworm* (yellowish-white with brown head, ¾ inch long)	Feeds on leaves, buds, flowers, stems, and eventually fruit		50% Sevin WP (2T to 1 gal. water)	Spray when symptoms of worms appear and weekly thereafter. Allow at least 1 day between last spray and harvest.
	Squash Vine Borer (white, up to 1 inch long)	Bores into vines; plants wilt and eventually die	See listing for STRIPED CUCUMBER BEETLE		
	Squash Bug (brown, flat back stinkbug, ⅝ inch long)	Colonies of bugs usually visible; leaves wilt from insect sucking out juices	Hand-pick beetles; destroy egg clusters; trap bugs under boards placed on soil near plants, collect and destroy each morning	50% Sevin WP (2T to 1 gal. water)	Spray when beetles appear and weekly thereafter until eliminated. Allow at least 1 day between last spray and harvest.
	Leaf Spots (fungus)	Small gray spots on leaves; plant wilts		80% Maneb WP (2T to 1 gal. water)	Spray plants when runners have formed.
Eggplant, Peppers, and Tomatoes	*Flea Beetle* (see listing under MOST CROPS)				
	Cutworm (see listing under MOST CROPS)				
	Tomato Hornworm (green with diagonal lines on side and prominent spike or "horn" on rear)	Chewed foliage and fruits	Hand-pick worms and immerse in kerosene	50% Sevin WP (2T to 1 gal. water)	Spray plants when hornworms appear.

EC Emulsifiable Concentrate (liquid)
WP Wettable Powder
T Tablespoon
t teaspoon

CROP	DISEASE/PEST	SYMPTOMS	CONTROLS		APPLICATION
			NONHAZARDOUS	SOME HAZARDS	
Eggplant, Peppers, and Tomatoes (*Cont'd.*)	*Leaf Miners* (see listing under MOST CROPS)				
	Blossom-end Rot	Dark, sunken, leathery spots at blossom end of fruit; likely to occur after dry spell during early growth	Apply dosages of lime and superphosphate to counter calcium deficiency; water garden evenly		
Leeks and Onions	*Onion Maggot* (white root maggot up to ⅓ inch long)	Shriveled useless bulbs caused by maggot borings		57% Malathion EC (2*t* to 1 gal. water)	Spray as soon as egg-laying flies are detected early in growing season. Repeat 3 times at 7-day intervals. Allow at least 7 days between last spray and harvest.
	Thrips (yellow to brown, winged and very active, about 1/25 inch long)	White blotches on leaves; tips of plants eventually wither and turn brown		57% Malathion EC (2*t* to 1 gal. water)	Spray entire plant at 7-day intervals to within 3 weeks of harvest.
Lettuce	*Leafhopper* (greenish-yellow, slender wedge-shaped, ⅛ inch long)	Sudden spread of virus diseases	Plant crop in sheltered area (leafhoppers prefer open spaces)	57% Malathion EC (2*t* to 1 gal. water)	Apply spray when plants are about ½ inch high and repeat weekly. Allow at least 7 days between last spray and harvest.
	Cabbage Looper (see listing under CABBAGE)				
	Drop (fungus)	Outer leaves wilt; watery rot on stems and old leaves	Provide good air circulation and drainage; hill soil slightly around plants to avoid water accumulation		
Muskmelon (*See Cucumbers*)					

			NONHAZARDOUS	SOME HAZARDS	
Onions (*See Leeks*)					
Parsnips (*See Carrots*)					
Peas	*Pea Weevil* (brown with gray, black, and white markings, 1/5 inch long)	Damaged blossoms and egg clusters on young pods		57% Malathion EC (2*t* to 1 gal. water)	Spray plants while blossoming and before initial pods appear. Allow 7 days between last spray and harvest.
	Seed maggot (see listing under BEANS)				
Peppers (*See Eggplant*)					
Potatoes	*Colorado Potato Beetle* (yellow with black stripes on wing segments, 3/8 inch long; larva pale red, soft-bodied, 3/5 inch long)	Chewed leaves and eventually defoliated plants; small plantings especially susceptible	Hand-pick beetles and larvae; crush egg masses when detected; check plants daily for effective control OR 5% Rotenone WP (5T to 1 gal. water)	50% Sevin WP (2T to 1 gal. water)	Apply when larvae are first detected. One or 2 sprayings usually sufficient.
Pumpkin (*See Cucumbers*)					
Radishes	*Root Maggots* (legless, yellowish-white, 1/3 inch long)	Tunneled, unusable roots		2% Diazinon granules (6 oz. per 100 sq. ft.)	One week before planting work granules into top 4 inches of soil.
Rhubarb	*Rhubarb Curculio* (yellow beetle with sucking snout, 3/4 inch long)	Small puncture marks in edible leaf stalks	Hand-pick beetles; remove and destroy all dock plants in vicinity of rhubarb planting where beetles breed		
Spinach	*Aphids*, *Leaf Miners* (see listing under MOST CROPS) *Cabbage Looper* (see listing under CABBAGE)				

EC Emulsifiable Concentrate (liquid)
WP Wettable Powder
T Tablespoon
t teaspoon

CROPS	DISEASE/PEST	SYMPTOMS	CONTROLS		APPLICATION
			NONHAZARDOUS	SOME HAZARDS	
Squash (*See Cucumbers*)					
Tomatoes (*See Eggplant*)					
Turnips (*See Carrots*)					
Watermelon (*See Cucumbers*)					
Blackberries and Raspberries	*Raspberry Crown Borer* (white, grublike larva, up to 1¼-inches long; hatches from oval, deep red eggs attached to undersides of leaves)	Burrows into bark at base of plants, later into crowns		25% Diazinon EC (2*t* to 1 gal. water)	Drench crowns and lower canes toward close of season. Repeat again 2 weeks later. **Do not apply Diazinon if fruit is present on plant.**
	Raspberry Fruitworm (beetles are yellow to brown, ¼-inch long; larvae are brown and white, ⅛-inch long)	Beetles chew long, narrow slits in blossoms and new leaves	5% Rotenone WP (5T to 1 gal. water) OR 2% Rotenone dust		Apply dust or spray to leaves 7 days after first blossoms appear and repeat 3 times at 10-day intervals.
	Rose Chafer (gray or fawn-colored beetle with reddish-brown head, slender and slow moving, ½-inch long)	Chews leaves, buds, flowers, and fruits	Hand-pick insect	50% Methoxychlor WP (2T to 1 gal. water)	Spray plants when chafers appear. Allow at least 3 days between last spraying and harvest.
Blueberries	*Mummy berry* (a fungus especially troublesome in damp, cool weather)	Shrunken, dried-out fruit; berries fall to the ground	Cultivate, hoe or rake area around plants once a week to discourage disease spore formation	80% Ferbam WP (3T to 1 gal. water)	Spray plants when buds are swollen and repeat every 7 to 10 days for 4 applications.
	Cherry Fruitworm (mature larvae are bright orange, about ⅜-inch long) *Cranberry Fruitworm* (small larvae have green undersides and brownish-red backs, ½-inch long)	Larvae attack green, unripened berries		50% Sevin WP (2T to 1 gal. water) OR 25% Malathion WP (2T to 1 gal. water)	Spray when most blossoms have separated from young berries. Repeat again in 7 to 10 days.

			NONHAZARDOUS	SOME HAZARDS	
Grapes	*Black Rot* (fungus)	Light brown circular spots on young leaves, then on berries; grapes eventually wither into hard dry mummies		80% Ferbam WP (3T to 1 gal. water)	Spray vines, especially new growth, at 10- to 20-day intervals until grapes are nearly full size.
	Downy Mildew (fungus)	Light yellow spots on older leaves developing into white moldy growths; leaves turn brown and fall off	Rake and cultivate lightly around plants to discourage growth of spores which can winter over	75% Zineb WP (2T to 1 gal. water)	From midseason on spray vines at 10-day intervals until just before berries ripen.
	Grape Leafhopper (pale yellow, jumping insect with red or yellow markings, ⅛-inch long)	Leaves mottled with numerous small white blotches	Remove garden waste near vines where insects can winter over	57% Malathion EC (2*t* to 1 gal. water) OR 25% Endosulfan EC (1T to 1 gal. water)	Spray vines once just after bloom and again 10 to 15 days later.
Raspberries	(*See Blackberries*)				
Strawberries	*Aphids* (see listing under MOST CROPS)				
	Cutworm (see listing under MOST CROPS)				
	Cyclamen Mite (smooth, whitish, and invisible to the naked eye)	Severe distortion, stunting, and discoloration of leaves; mites feed on blossoms causing misshapen fruit	Set out only pretreated, uninfested plants; *dormant* plants should be immersed in water heated to 100 degrees	25% Endosulfan EC (1T to 1 gal. water)	Apply spray over entire bed. Allow at least 4 days between last application and harvest.

EC Emulsifiable Concentrate (liquid)
WP Wettable Powder
T Tablespoon
t teaspoon

II.

Individual Crops -

A Practical Guide To Growing Them

6. Vegetables

When it comes time to match seed to garden plan, your best ally may turn out to be a seed catalog from a reliable seedsman that presents a gallery of vegetables and fruits in all their glory of full ripeness. Although the full-color pictures might easily prompt you to bite off more than you can either chew or grow, the information accompanying the illustrations is well worth reading. Look for disease-resistant varieties. Consider the length of the growing season until harvest, bush versus pole varieties, and hybridized new versions of old standards. Tomatoes, for example, range from a pastel pink to deep red, not to mention the yellow types. Size is also a factor to consider; there are dwarf peas and cherry tomatoes, and there are extra-long cucumbers—almost 20 inches in length. And don't hesitate to experiment with unknowns. Every garden should have a "fun row" of spaghetti squash, peanuts, soybeans, or gourds. Keep the children in mind too. The giant Hubbard squash, sunflower, or pumpkin can provide any youngster with a Jack-and-the-beanstalk world right in his own backyard.

To help you find your way among the maze of requirements and special planting needs of the various vegetables and fruits, we have arranged them in alphabetical order, vegetables in this chapter and small fruits in the next. You'll find the step-by-step instructions for each in a logical, start-to-finish order: *planting, care and feeding, harvesting,* and *insects and diseases.* At the top, along with the name of the vegetable or fruit and the varieties, we have included a shorthand summary of the critical things you should know when planning your garden, putting together a seed order, and, finally, planting. These are the headings and what they mean:

Variety The term "variety" is nothing more than a handy way to talk about different plants of the same vegetable crop. The wide choice of varieties of most vegetables available from seed catalogs is what makes gardening full of delightful surprises. The varieties are the result of constant crossbreeding and hybridizing, which are done to enhance size, sweetness, disease resistance, hardiness, and other desirable characteristics. Each variety of vegetable possesses distinctive qualities. Golden Bantam corn, for example, is noted for large, yellow kernels while Silver Queen corn has large, clear, white kernels. The variety name given is often descriptive (Big Boy tomatoes, Sugar Baby watermelons) or the name may be derived from the location or the breeder himself (Connecticut Field pumpkins, Texas Grano onions, Emerson Pascal celery, Burpeeana cauliflower).

When shopping through your seed catalog be sure to order true F_1 seeds —that is, first generation seeds.

Hardiness Vegetables fall into two broad categories, according to their ability to resist low or freezing temperatures. This is a factor to reckon with in most areas, though in warm or southerly regions it is less likely to be vital. *Tender* crops are susceptible to frosts at both ends of the season, while *hardy* varieties are able to withstand light frosts.

Time to Plant Exactly when to plant depends on a crop's resistance to cold and, in some cases, to heat. Another factor to consider is the time it takes to mature to the harvesting stage, which may tell whether or not succession planting will be possible or—in the extreme north—whether the season is long enough to bring it to maturity. In any case, hardy crops can be planted before the average date of the last frost in your area, while tender crops need a warm soil and frost-free weather before it is safe to plant them outdoors. For the probable frost-free dates in your area, consult the appendix on page 218.

Spacing beween Rows and Plants All recommended distances between rows and plants are stated in inches. Maximum and minimum figures are given to allow you to provide for either machine (maximum) or hand (minimum) cultivation. Machines require more room, of course, which means rows will have to be spaced wider apart.

Depth to Plant Seed The depths indicated are meant as rough guides only. Remember that, as you dig deeper into the soil, the moisture content increases and the temperature decreases. Seeds requiring a warm soil in spring should be planted near the surface. As a general rule, plant seeds to a depth equal to about four times their diameter.

Seeds or Plants per 25 Feet of Row Most seeds available commercially come in small paper packets containing enough seed for a row 20 to 30 feet long. Farmers' supply outlets and seed catalogs offer quantities ranging from a packet to several pounds. Check the package or seed catalog carefully to determine your exact needs. Plants started in a commercial greenhouse are more expensive, so be sure to calculate accurately the amounts you intend to grow before you buy.

Amount per Person All we can do here is provide a rough estimate, since tastes and eating habits vary. If, however, you plan to freeze, store, or can your produce, or perhaps give or sell some to your neighbors, you might consider increasing the recommended amounts. But if you're just starting out, follow these guidelines fairly carefully—nothing dampens the gardening spirit quicker than too little or too much.

Yield per 25 Feet of Row The quantity of produce you realize from your gardening efforts obviously depends a great deal on locale, weather conditions, and the kind of care given to crops. The yield figures here represent what can be expected under average conditions.

Time to Harvest The number of days required from planting until a crop can be harvested depends on the growing habits of different varieties of the same vegetable or fruit. Seedsmen and Agricultural Experiment Stations are continually striving to develop quicker-maturing varieties. Seed catalogs usually list the number of days required to mature the crop. Study the seed catalog in order to match the variety and its growing season to your particular needs.

ASPARAGUS

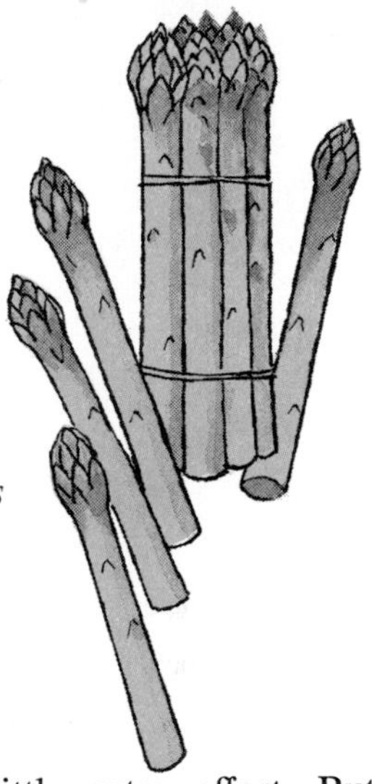

Varieties
Washington Strains

HARDINESS: *Hardy perennial*
TIME TO PLANT: *6 weeks before frost-free date*
SPACING: *rows 60 inches*
plants 18 inches
DEPTH TO PLANT ROOTS: *6 to 8 inches*
PLANTS PER 25′: *12 plants*
AMOUNT PER PERSON: *15 feet*
YIELD PER 25′: *12 pounds*
TIME TO HARVEST: *2 years*

At the outset asparagus unquestionably requires a little extra effort. But once healthy one-year-old crowns have been established the asparagus bed will reward the gardener with succulent spears every spring for as long as 15 years. In the long run it will turn out to be the least demanding but most rewarding crop in the family garden.

Planting Although asparagus can be started from seed, it makes much more sense to obtain well-established crowns no more than one year old from a reliable nursery. For the average family of four, anywhere from 30 to 50 plants should be sufficient, but if freezing or canning is planned then the bed should be expanded. Try to resist the offering of a well-intentioned neighbor who arrives at your doorstep with plants culled from his own patch. Since male and female flowers are produced on separate plants, there is the possibility of unwanted crosses and you might find yourself fussing over inferior plants. Also, you may discover other plants are not rust-resistant.

Asparagus likes cold winters, making it more suitable for northern regions than for southern regions, where yields will be much lower. The best soil is a deep sandy loam, but asparagus will do well on almost any soil providing it is well drained and rich in nutrients, and not below pH 6.0. Humus, either in the form of green manures plowed under or organic matter such as compost or manure, is an important prerequisite. In spring, as soon as the soil can be worked, dig furrows or trenches 18 inches wide and 14 to 16 inches deep. Mix generous amounts of organic matter—compost, peat moss, etc.—and a complete fertilizer (five pounds of a 5–10–10 fertilizer per 75 feet of row) with the removed soil and fill the trench to within 8 to 10 inches of ground level. Place asparagus crowns 18 inches apart and cover with several inches of soil. Then add more soil as the plants develop. By the end of the season the trenches should be all but invisible.

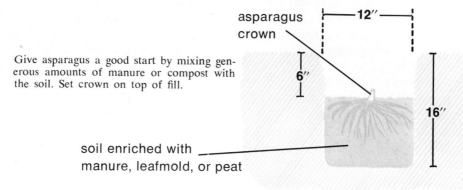

Give asparagus a good start by mixing generous amounts of manure or compost with the soil. Set crown on top of fill.

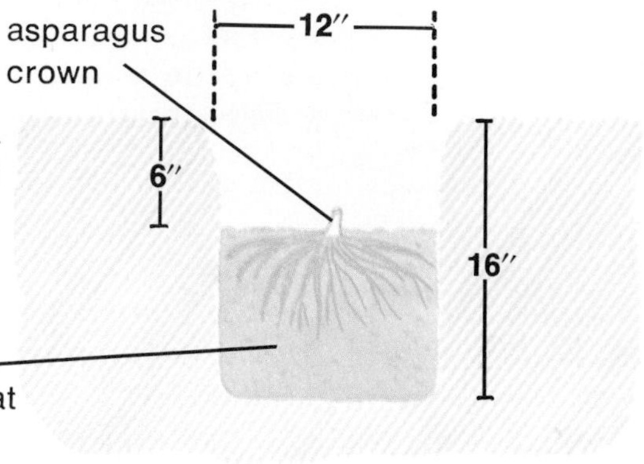

Care and Feeding A good way to outfox weeds is to use the plot intended for the asparagus bed the year before to grow a root crop that requires frequent cultivation. Perennial weeds will get the hint and will present little or no problem the following spring when the asparagus crowns are set out. After that, except for an occasional hoeing the bed will require little attention. Be sure to keep cultivation shallow to avoid injury to roots. Remember, too, that as the plants mature their roots will grow toward the surface.

Since asparagus is a heavy feeder, soil fertility should be maintained by frequent applications of a commercial fertilizer. Simply broadcast about 15 pounds of a 10–10–10 fertilizer per 1,000 square feet just after the first cultivation and another 15 pounds after the last cutting. Or in the fall, after the tops have been cut off at ground level, shredded, and returned to the soil (or disposed of), mulch the bed with a thick layer of manure or compost. Other mulches excellent for thwarting weeds are pine needles, sawdust, wood chips, peat moss, or ground corncobs. If wood chips, shavings, or sawdust are used, an occasional application of a nitrogen fertilizer will be necessary to avoid a nitrogen deficiency in the soil.

Harvesting The ideal asparagus spear should be tender, not too thick, with a compact tip. Wait until the spears are about eight inches in height, and then cut one-half to one inch below the surface of the soil. When cutting, hold the knife vertically to prevent injury to sprouts that have not yet broken through the soil.

Put the pot on to boil before you go out to harvest. Like corn and peas, asparagus is at the peak of flavor when cooked immediately.

A new bed should not have spears cut until the third year after planting. However, a few spears may be removed the second year provided the cutting period is kept short (two or three weeks at the most). In subsequent years all spears can be cut until July 10, then allow the tops to grow and mature into full-fledged fern-leaved plants. Once the tops have been nipped by a stiff frost, cut them, shred and add to the compost heap, or dispose of them otherwise.

Insects and Diseases The spotted asparagus beetle and the common asparagus beetle (metallic blue to black with orange or yellow markings and about one-fourth inch long) are common invaders of the bed. Adult beetles gnaw on plants and shoots, while the larvae often attack the crowns themselves and also the berries, causing reduced yields. Since adult beetles manage to overwinter under garden litter, it is important that garden refuse be disposed of. In moist weather asparagus rust may be a problem. Symptoms include orange-red powdery blisters on stems and foliage. The best cure is prevention—grow a rust-resistant variety such as Mary Washington. For specific diseases and controls see the chart in chapter 5.

BEANS, GREEN or SNAP

Varieties
BUSH
Contender
Resistant Cherokee (wax)
Tender Crop
POLE
Kentucky Wonder

HARDINESS: *Tender*
TIME TO PLANT: *After danger of frost*
SPACING: *rows 24 to 30 inches*
plants 2 to 3 inches
DEPTH TO PLANT SEED: *1 to 2 inches*
SEEDS PER 25′: *¼ pound*
AMOUNT PER PERSON: *10 to 15 feet*
YIELD PER 25′: *11 to 12 pounds*
TIME TO HARVEST: *50 to 60 days*

Like most other beans, snap beans are quite cold-sensitive and cannot be planted until after all danger of frost is past. Two types are available, green and yellow, both of which will grow in a wide range of soils. Both are available as pole beans, too. If you have a congenital weakness for beans you might enjoy sifting through seed catalogs for more exotic bean varieties such as dry or kidney beans, the Fava or broad Windsor, or even some of the edible soybeans. Whatever your choice, plan on succession plantings in order to enjoy a bountiful and continuous supply. Plantings spaced seven to ten days apart will keep the dinner table well supplied through the season.

Planting The cultural requirements of the snap bean are similar to those of the lima bean. Since the seed reacts badly to damp cool conditions, the soil should be well drained and thoroughly warmed before you attempt to plant. To prevent soil crusting, cover the seed with peat moss and sand. Two 25-foot rows spaced 24 to 30 inches apart should be just about the right amount for a first planting for the average-size garden. Pole beans can be 36 inches apart in the row. Remember to reserve space for succession plantings as the summer wears on. In the Deep South and Southwest, beans can also be grown during the fall, winter, and spring.

Care and Feeding Beans will thrive on any moderately fertile soil and do not necessarily require additional feeding, although side applications of a complete fertilizer raked into the soil between the rows will do no harm, particularly in light sandy soils. When broadcasting fertilizer just before planting, be sure it does not come into direct contact with injury-susceptible seeds. Use three to four pounds of a 5–10–10 or 4–8–4 commercial fertilizer for every 100 square feet of garden area. Incorporate granules into the top four inches of topsoil. Bloodmeal, cottonseed meal, phosphate rock, and wood ashes are excellent nutrient-rich materials if you prefer the organic approach.

Harvesting A chief advantage of snap beans is that they can be picked at almost any stage of their development. Since there is no peak point of ripeness, the harvesting season is considerably extended. You'll find that beans on the young side are much less stringy than overripe beans. Many varieties are offered that are practically stringless when picked at the right time.

Insects and Diseases Remove the spent vines after harvest in order to prevent insect and disease problems next time you plant. For specific pests, diseases, and controls, consult the chart in chapter 5.

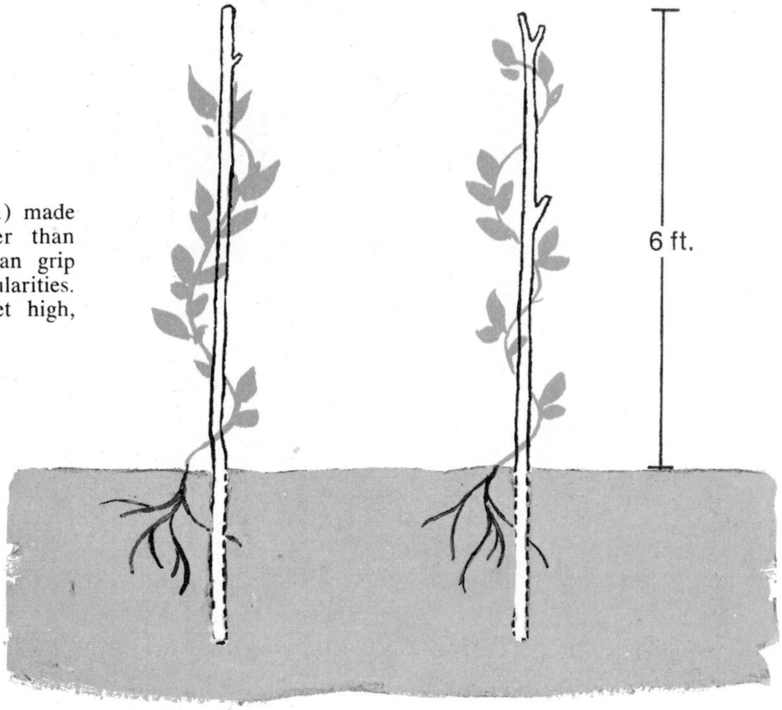

Bean poles (for both snap and lima) made from saplings are generally better than lumberyard stakes because vines can grip the rough bark and other irregularities. Poles should be from 6 to 8 feet high, driven 2 feet into the soil.

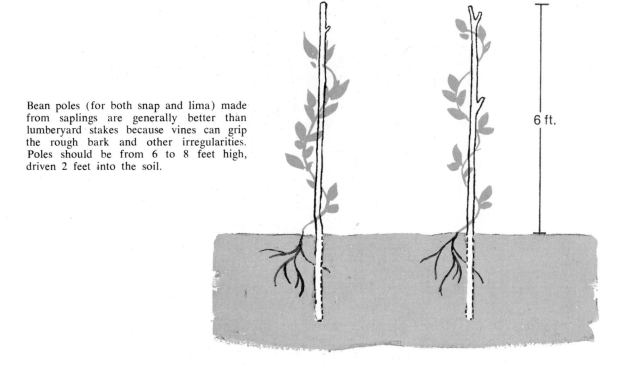

Use wire and string method for longer rows.

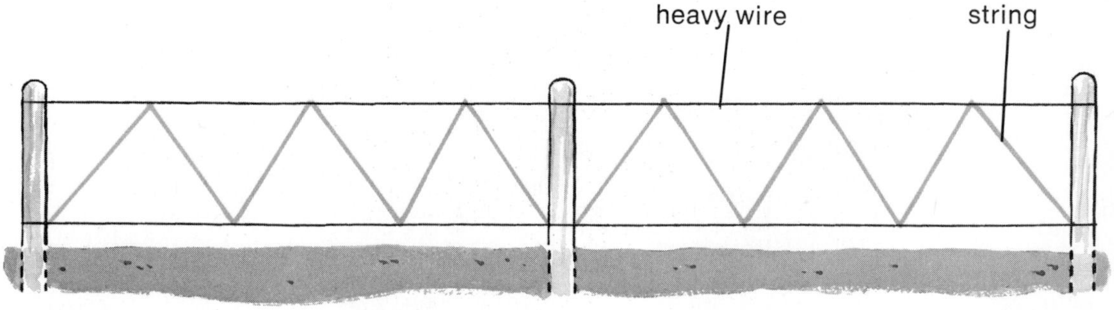

BEANS, LIMA

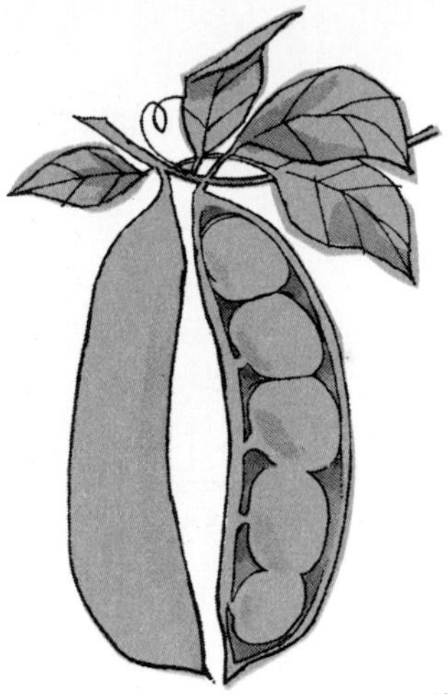

Varieties
Fordhook 242
Henderson

HARDINESS: *Tender*
TIME TO PLANT: *After danger of frost*
SPACING: *rows 26 to 48 inches*
plants 12 inches
DEPTH TO PLANT SEED: *1½ inches*
SEEDS PER 25′: *¼ pound*
AMOUNT PER PERSON: *40 feet*
YIELD PER 25′: *15 pounds*
TIME TO HARVEST: *65 to 75 days*

Lima beans are a decidedly warm-weather crop and need a gentle climate for good yields. A cold moist soil will rot the soft-skinned seeds and cool nights will retard growth, so be sure the soil is thoroughly drained and warm before planting. Bush limas are more successful in cooler sections than pole varieties.

Planting Although there is little appreciable difference between the low-growing bush variety of lima bean and the taller-growing pole variety, cultural requirements are somewhat different.

BUSH LIMAS—After the soil has been thoroughly prepared, mark the row with garden twine and with a hoe dig a furrow about one to one-and-a-half inches deep. Space the seeds from six to eight inches apart and cover with an inch of soil. Tamp down the soil with the flat side of the hoe to be certain that air pockets are not created. When seedlings emerge and the plants have become fairly well established (about four inches high), thin the row leaving 12 inches between plants. If the soil looks as if it might crust and imprison the seed shoots, spread a mixture of sand and peat moss (about half and half) over the seeds at planting time instead of regular soil.

POLE LIMAS—Pole beans, because they grow up instead of sidewise, offer a handy way to make the most of cramped space. If a minigarden is all you can manage, try growing beans along a fence or alongside a porch. The important thing for them is plenty of sun and warm nights.

The pole variety can be grown in either rows or hills. For hills, gather soil into mounds spaced four feet apart. Plant four to five seeds in each hill, and later thin to three plants per hill. For row planting, place the seeds four inches apart in a furrow about two inches deep. When plants reach a height of from eight to ten inches, support will have to be provided. Cut wood saplings six to seven feet high; they make excellent supports because the rough bark surface gives the winding vine something to cling to.

Care and Feeding If the soil has been prepared with ample amounts of manure or fertilizer, the chances are no additional applications will be necessary. But beans, like any other vegetable or fruit, will not appreciate too much competition from unruly weeds. Cultivate frequently or use a mulch to smother weeds. Be sure to stay out of the bean patch when plants are wet, since disease can spread rapidly as the careless gardener brushes past wet leaves.

Harvesting The perfect bean pod should feel plump and somewhat firm to the touch. See if you can catch the lima on the tender side, just before peak ripeness. A bean allowed to remain on the vine too long will be mealy and pulpy.

Insects and Diseases The lima's most widespread pest is the Mexican bean beetle, which looks dangerously similar to the beneficial ladybug. Before reaching for a spray can, look for the bean beetle's slightly larger size, its copper yellow—not orange—color, and 16 spots (not 12) that appear in three distinct rows from left to right across the wing segments. The best preventive is to remove attractive nesting sites such as weeds and garden refuse. See the chart in chapter 5 for other pests and their remedies.

BEETS

Varieties
Crosby
Early Wonder
Detroit Dark Red

HARDINESS: *Hardy*
TIME TO PLANT: *When soil can be worked*
SPACING: *rows 14 to 24 inches*
plants 2 to 3 inches
DEPTH TO PLANT SEED: *½ to 1 inch*
SEEDS PER 25′: *1 packet*
AMOUNT PER PERSON: *10 to 15 feet*
YIELD PER 25′: *18 pounds*
TIME TO HARVEST: *60 to 70 days*

Beets not only have a delicious flavor but they can also be used as a kind of thermometer of the soil pH level. They are sensitive to acid soils, and a sickly stand of beets indicates some kind of liming program is needed. See chapter 2 for ways to "sweeten" an acid soil and the amount of material to use.

Planting Beets generally do poorly in hot weather, so fall, winter, and spring plantings are recommended in southern areas. In northern regions succession plantings every two weeks will provide a continuous supply. Quality beets depend on quick maturation. To encourage rapid growth, work the soil thoroughly, removing stones, twigs, and extraneous trash.

With garden twine mark off rows 14 to 24 inches apart, being certain to reserve space for a row or two of later plantings. Dig a shallow furrow no deeper than one inch by dragging a hoe handle along the taut twine. Drop the seeds individually at the rate of about six seeds per foot. If the soil is heavy or tends to form a hard impenetrable surface after a light rain, a mixture of peat moss and sand or leaf mold should be used to cover the seed. A sprinkling of water will provide the necessary moisture to trigger the germination process. After seedlings have emerged and appear sturdy, thin the plants to four or six plants per foot.

Care and Feeding Beets, while sensitive to acid soil, respond well to generous applications of a 5–10–10 fertilizer, three pounds per 100-foot row before and after planting. Wood ashes are also good for soils where beets grow. But truckloads of fertilizer aren't going to help much if the soil is basically poor in organic content. Try to equal the fertilizer dosage with at least equal amounts of organic matter, in the form of either well-rotted manure or compost produced from last year's garden refuse and lawn clippings. Since water is an indispensable element in the nutrient-exchange process, frequent waterings are always in order but should not exceed approximately one to one-and-a-half inches per week, the average requirement for most vegetables and fruits.

Weeds, of course, are never welcome. Use a wheel hoe or common hoe to eliminate weeds from midrow. Weeds invading the row itself can only be dispatched by hand. If either machine or hand cultivating brings phantom pains to the lower back, using a mulch of some kind will be easier on you physically and will accomplish the same result. Straw, lawn clippings, or sawdust creates a neat appearance, smothers weeds, and preserves moisture as well. Be certain that the mulch doesn't become too thick and block air circulation. When settled, a good mulch shouldn't be more than two inches thick.

Harvesting As beets reach peak maturity the root tops will begin to protrude through the soil. When the beets look as though they are an inch and a half to two inches in diameter they are ready for pulling. But don't forget the delicious possibilities in the greens. Prepared like spinach, the edible tops make beets an especially desirable crop for the home garden. Make plantings thicker than needed so thinned-out plants can be prepared for the dinner table.

Insects and Diseases If the soil is adequately prepared, treated seed used, and the garden kept free of refuse and weeds, the beet crop should be fairly free of insects and disease. Where winters are severe enough to kill off lingering bugs and larvae, beets are almost disease-free and one of the easiest crops to grow. In other areas the beet leafhopper, webworm, and blister beetle may be a problem. For specific instructions, see the chart in chapter 5.

BROCCOLI

Varieties
Green Sprouting
Waltham 29

HARDINESS: *Hardy*
TIME TO PLANT: *When soil can be worked or indoors*
SPACING: *rows 30 to 36 inches*
plants 16 to 22 inches
DEPTH TO PLANT SEED: *½ to 1 inch*
SEEDS/PLANTS PER 25′: *1 packet/15 plants*
AMOUNT PER PERSON: *5 plants*
YIELD PER 25′: *12 pounds*
TIME TO HARVEST: *40 to 85 days*

Grown for its tender blue-green flowerets, vitamin-rich broccoli is resistant to cold and will thrive on any fertile soil as long as plenty of moisture is available.

Planting Since the broccoli planting should be timed so that flowerets mature during the cooler portions of the growing season, it's best to get seed started indoors.

About four weeks before the earliest outdoor planting date for your area, sow seeds in flats containing two to three inches of soil topped with sterile sphagnum moss or vermiculite. Saturate with water and cover with newspaper or cardboard to keep out the light. A pane of glass placed over the flat before the newspaper is added is an excellent way to keep in moisture. When the seeds have sprouted, pinch out excess seedlings to prevent overcrowding. To prepare broccoli for the shift to outside conditions, place in a cold frame for at least a week. Then, when soil conditions permit, set plants in the garden 18 to 24 inches apart. See chapter 3 for hints on successful indoor and outdoor planting.

Care and Feeding In addition to ample water, broccoli will respond well to cow, horse, or chicken manure worked into the soil. Remember, though, that fresh manure is potent and can easily burn roots if not thoroughly mixed with the garden soil at least two weeks before plants are set out. Or use a high-nitrogen commercial fertilizer (10–6–4), spread at the rate of five pounds for every 100 square feet. For good aeration and water movement through the soil, cultivation should begin immediately. Be sure to keep hoeings shallow near the row to avoid injury to surface roots.

Harvesting Broccoli as it appears on the dinner table is essentially a bud, which should give some clue as to when it should be harvested. Try to get sprouts before the tiny green flowerets have begun to open. With a kitchen knife, slice off the top of the center portion—two to four inches of stem, not too close to the head. Leave the remaining side shoots for later pickings.

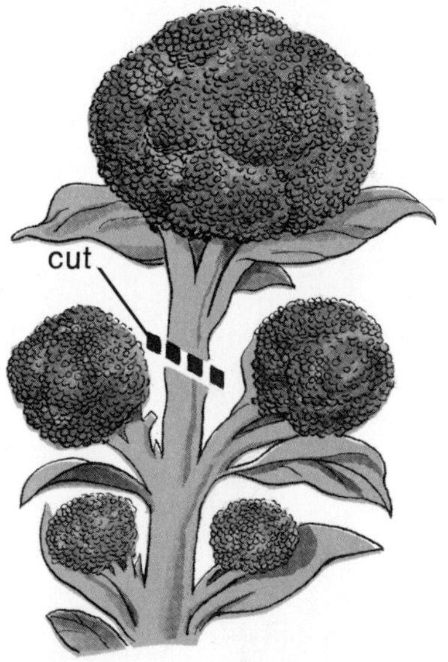

When harvesting broccoli cut off only the uppermost "flower" cluster, leaving side shoots to develop for later cuttings.

Insects and Diseases The greatest enemies of broccoli are club root, black-leg, and black rot. These soil-borne diseases can be avoided by rotating all the crops of the Brassica family, which include cabbage, cauliflower, and Brussels sprouts as well as broccoli. Occasionally a cabbage looper might be detected chewing on broccoli leaves. Either hand-pick or spray with rotenone or malathion. See the chart in chapter 5 for other pests as well as pesticide precautions.

BRUSSELS SPROUTS

Varieties
Jade Cross

HARDINESS: *Very hardy*
TIME TO PLANT: *When soil can be worked or indoors*
SPACING: *rows 24 to 30 inches*
plants 18 inches
DEPTH TO PLANT SEED: *½ to 1 inch*
SEEDS/PLANTS PER 25′: *1 packet/50 plants*
AMOUNT PER PERSON: *5 plants*
YIELD PER 25′: *12 pounds*
TIME TO HARVEST: *80 to 100 days*

Brussels sprouts do best under relatively chilly conditions and, in fact, will improve in flavor after a mild frost. Also a member of the Brassica family, Brussels sprouts are similar to broccoli in their cultivation needs and, like broccoli, reward the gardener with a continuous yield of tasty, leafy buds that have no equivalent when it comes to uniqueness of flavor. Because of this plant's cool-weather habits, it can be grown as either an early-season or a late-season crop. In southern sections it will easily winter over, though it may not produce as well as in the North.

Planting The Brussels sprouts crop really gets under way indoors where plants are started four to five weeks before their early-spring planting outdoors in order to gain time. Spread the seed evenly in a flat of sterile soil and cover to preserve moisture. The ideal temperature for most seeds to germinate is in the neighborhood of 70 degrees. When seedlings have appeared and unwanted plants are thinned out, place the flat in sunlight and keep soil moist. Once hardened off in the cold frame, the plants can be placed in the garden 18 inches apart in rows approximately 24 to 30 inches apart. To ward off cutworms, put an aluminum foil or cardboard collar around the stem of each plant when setting out (see page 44).

Care and Feeding A reliable supply of water and an occasional loosening of the soil with a hoe are all the maintenance Brussels sprouts will need. To keep the soil in good rain-receiving condition, cultivate around the plants with shallow strokes of the hoe. If things are getting somewhat dusty underfoot and a drought seems imminent, begin a regular watering program, especially when plants are young. Because of their sluggish growing pace, requiring nearly four months before the buds can be picked, Brussels sprouts are a good candidate for intercropping with more rapidly developing crops. Planting radishes, lettuce, or spinach within the row makes maximum use of space and utilizes soil that might otherwise be given by default to weeds. In the early stages of growth a side dressing of high-nitrogen fertilizer will be welcome.

Harvesting The prized morsels of Brussels sprouts are the small globe-shaped buds couched in the joints where leaf meets stem. As they grow, their demands for space will increase. At this point the lower leaves should be gently broken off by twisting them away from the stalk. Sooner or later the plant will look like a shrunken coconut tree with just a tuft of leaves sprouting from a naked-looking stem. Be sure to allow some leaves to remain in order for the plant to continue functioning. In terms of harvesting, timing is what will make the difference between delicate-tasting sprouts and hard, tough, tasteless buds. If the lower leaves have turned yellow you are too late. At the same time, the small buds resembling miniature cabbages should be at least an inch in diameter and not changing in color. They can either be snapped off the stem or removed with a kitchen knife. If it looks as if winter might abbreviate the harvest, dig up the plants, place them in the cellar in boxes or in a cold frame, and tamp soil around the roots. Continue to pick developing sprouts. In milder areas, of course, the plants can be left in the ground throughout the winter.

Insects and Diseases Brussels sprouts are prone to the same soil-borne diseases as its sister crops of cauliflower, cabbage, broccoli, and kale. Consequently, Brussels sprouts should be given a change of venue every season by rotating crops. For symptoms and control of common pests, consult the chart in chapter 5.

CABBAGE

Varieties
Resistant Golden Acre
Copenhagen Market
Resistant Danish
Resistant Red Acre
Chieftain Savoy

HARDINESS: *Hardy*
TIME TO PLANT: *When soil can be worked or indoors*
SPACING: *rows 24 to 36 inches*
plants 14 to 24 inches
DEPTH TO PLANT SEED: *½ inch*
SEEDS/PLANTS PER 25′: *1 packet / 13 plants*
AMOUNT PER PERSON: *5 plants*
YIELD PER 25′: *32 pounds*
TIME TO HARVEST: *60 to 110 days*

Adapted to almost all soils and very resistant to cold, cabbage turns up in more vegetable patches than perhaps any other vegetable. A well-nourished, friable soil kept reasonably moist is the principal requirement.

Planting Start the seed in flats or pots indoors long before the last patches of snow have melted (see chapter 3). For an average family of four you might plan on at least five plants per person, but be certain you sow extra seeds to make up for plants that may fail. When transplanting, seedlings can be planted deeper than the previous soil level. Solid, low-growing heads will result. But cabbage, like other crops started indoors, needs a short adjustment period in the cold frame. Plants already started may be bought but with less choice of varieties.

Care and Feeding A heavy feeder that improves in quality with increased rate of growth, cabbage should receive healthy dosages of manure or commercially prepared fertilizers. Manure is especially favored because it supplies some nutrients and at the same time improves the structure of the soil. The better the soil structure, the more moisture is made available to roots. A nitrate of soda fertilizer can also be used to good advantage a month after the plants have become established. Spread it by hand along the row at the rate of one pound per 100 feet of row, or one-third ounce for every plant. Or sprinkle an organic high-nitrogen material such as dried blood between the rows. Without water, though, all the fertilizer this side of China is not going to do much good. Cabbage needs plenty of moisture so water frequently, especially during dry spells or in dry areas. A light mulch of straw, hay, or dried lawn clippings will thwart weeds and conserve moisture.

Harvesting When the heads are solid and have not begun to crack, they are ready for picking. If the head starts to crack before reaching full size, try cutting the roots on one side of the plant with a spade when the soil is moist. Cabbage, like many other vegetables, is best when tossed into the pot right away. A cabbage crop gathered in the fall should be stored at near-freezing temperatures and in relatively high humidity. Use only well-formed, injury-free heads for storage.

Insects and Diseases The leading member of the Brassica family, cabbage should be rotated to a new site yearly to avoid yellow blight, root-knot, and other ills. Yellow blight, caused by bacteria that enter the plant through its roots, makes leaves wilt and change color. Look for a resistant variety when ordering seed. The cabbage looper and aphid often search out cabbage. Both can be controlled with nontoxic sprays recommended in chapter 5.

CARROTS

Varieties
Nantes
Red Cored Chantenay

HARDINESS: *Hardy*
TIME TO PLANT: *When soil can be worked*
SPACING: *rows 16 to 24 inches*
plants 1 inch
DEPTH TO PLANT SEED: *½ inch*
SEEDS PER 25′: *1 packet*
AMOUNT PER PERSON: *10 feet*
YIELD PER 25′: *25 pounds*
TIME TO HARVEST: *70 to 75 days*

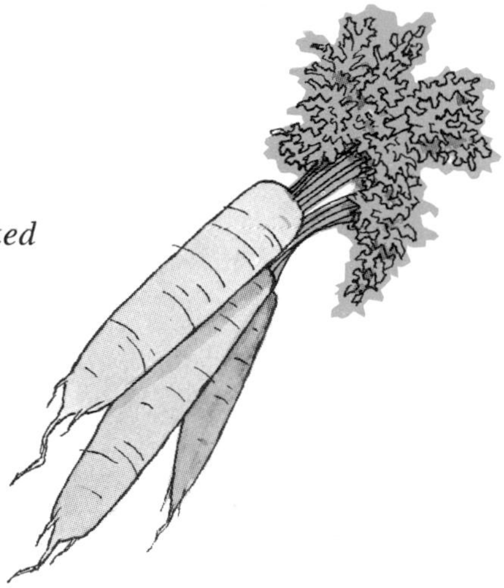

Of all the vegetables, few offer the rewards in store for you from setting out a row or two of carrots. They are cold resistant, easy to grow, and adapted to almost any kind of soil.

Planting Often slow in germinating, carrots can be sown as soon as the soil will cooperate with the spading fork. Mark the row with garden twine and dig a shallow furrow by dragging the handle of the hoe or other garden tool along the taut line. Carrot seeds take their time sprouting, so don't panic if the row begins to look like a wasteland of young weeds. In order to distinguish weed from carrot sprout, try mixing some rapidly germinating radish seed right in the packet of carrot seed before you plant. The emerging radishes not only will mark the row but also can be pulled to provide an early treat for the dinner table. Thin carrot seedlings to about 10 to 15 plants per foot.

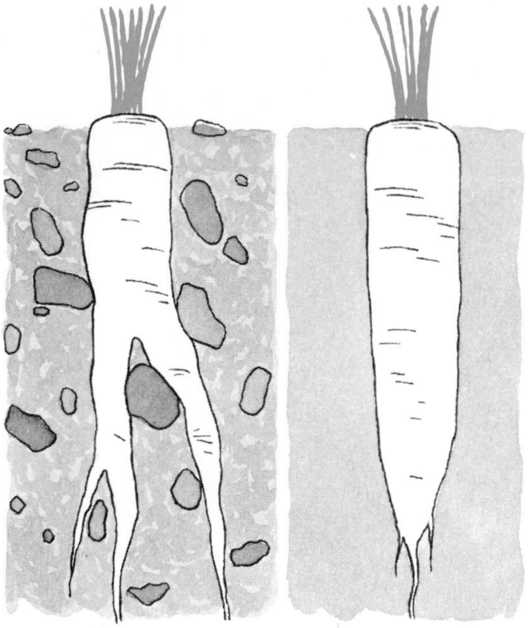

Stones and other trash, if not removed by careful raking, will cause forked or misshapen carrot roots. A well-prepared soil will produce attractively shaped carrots.

Remember that all root crops need a fine, well-prepared soil. Loosen the soil to a depth of ten inches and remove stones and other large debris with a steel rake. If your layer of dark topsoil appears shallow, look for a stubby variety of carrot such as Commander or Short n' Sweet, which produce thicker and shorter roots. Plant succession crops every few weeks till August.

Care and Feeding To encourage germination, gently water the row after the soil has been lightly tamped over the seed. Wait until the radish seed sprouts before attacking weeds. Then either remove them by hand or cultivate between rows using a shallow scraping motion with the hoe.

Harvesting That maxim "the biggest isn't the bestest" applies more to carrots than to any other crop. Carrots should be pulled before they are even half grown. Pick when the root tops are barely visible above ground, about an inch in diameter. Serve immediately or process for freezing. When stored under cool moist conditions, carrots can keep their freshness for as long as four months.

Insects and Diseases Normally carrots will not be molested by any insects or diseases. Carrot caterpillars might be found chewing on the tops, but they are seldom numerous enough to threaten the crop and can be easily removed by hand. Other possible troubles include leafhoppers, wireworms, leaf blight, and yellows. For specific antidotes see the chart in chapter 5.

CAULIFLOWER

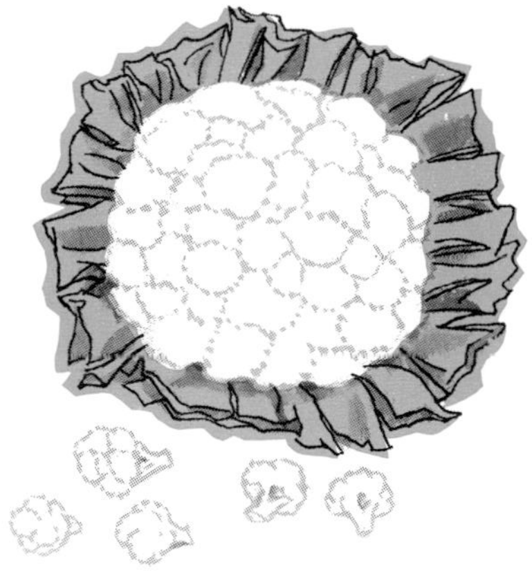

Varieties
Burpeeana
Purple Head
Snowball Types

HARDINESS: *Hardy*
TIME TO PLANT: *When soil can be worked*
SPACING: *rows 24 to 30 inches*
plants 20 to 24 inches
DEPTH TO PLANT SEED: *½ inch*
SEEDS/PLANTS PER 25′: *1 packet/14 plants*
AMOUNT PER PERSON: *5 plants*
YIELD PER 25′: *20 pounds*
TIME TO HARVEST: *60 to 90 days*

Most authorities consider cauliflower too ornery a vegetable for the limited patience of the beginning gardener. But if the family is crazy about fresh cauliflower, forget the experts and give it a try. The fact is, cauliflower needs no more extra attention than the ever popular tomato.

Planting Since cauliflower does best in cool temperatures, seeds should be started indoors. If you are planning on five plants per person for a family of three, sow enough seed so that 20 to 25 seedlings appear. When the plants are three inches high, transplant the strongest to individual peat pots or modified milk cartons. After a brief hardening-off period in the cold frame, set them out in the garden. Keep watered and free of advancing weeds.

Care and Feeding One or two side dressings of nitrate of soda or a high-nitrogen organic material will encourage growth and prevent premature "bolting" (going to seed). But the main secret to a good crop is lots of water. Give the cauliflower patch a thorough soaking at least once a week so that plants get one to one-and-a-half inches of moisture every seven days. From here on it's clear sailing until the small white center buds begin to appear. Since sunlight striking the bud can cause discoloration and sometimes a disagreeable flavor, you should gather together the lower leaves and tie them up over the plant with soft twine or a rubber band to blanch. If blanching strikes you as too time consuming, try the Purple Head variety, which tastes the same and doesn't need tying.

Tie cauliflower leaves up over the maturing flower bud to insure a rounded, pure white, and tasty head.

Harvesting Inspect every day or two in order to catch the heads at their prime. The ideal cauliflower head is about six inches in diameter, compact, and bone white in color. If in doubt, it is better to cut heads early than risk getting them too late. When cooking, steam rather than boil to retain full flavor, or boil briefly. Do not overcook.

Insects and Diseases Cauliflower is susceptible to the same variety of insects and diseases as cabbage and other Brassicas. The chief worry will be the soil-borne diseases that attack the plant through its roots. Resistant varieties and rotation planting are the best control methods. See also the chart in chapter 5.

CELERY

Varieties
Utah 52–70
Emerson Pascal

HARDINESS: *Hardy*
TIME TO PLANT: *Early spring or midsummer*
SPACING: *rows 24 to 36 inches*
plants 6 to 10 inches
DEPTH TO PLANT SEED: *½ inch*
SEEDS/PLANTS PER 25′: *1 packet/35 plants*
AMOUNT PER PERSON: *5 plants (or more)*
YIELD PER 25′: *35 pounds*
TIME TO HARVEST: *115 to 130 days*

Although a prima donna among vegetables because of the extra attention it needs, celery has a unique flavor that makes it an attractive crop for the home garden. For the cook, celery is practically indispensable. The tops, the strongest-flavored portion of the plant, are ideal for soups; the stalks make a delicately flavored cooked vegetable; and the hearts are perfect for the relish tray. Very low on the calorie list too.

Planting Celery should be given a head start indoors or in a cold frame, since it is a long-season crop. Although a cool-weather crop, celery needs approximately ten weeks to grow good plants. The seeds are small, hard, and somewhat testy. To soften tough seed coverings, put the seed in a cloth bag and soak in water overnight. Then plant in soil to a depth of one-fourth inch and cover with a thin layer of sand. Keep the soil and seed moist by covering with burlap that is kept dampened. A highly fertile soil with plenty of water is needed. Celery does not do well in hot dry climates.

Once the delicate seedlings have emerged, it is important to keep a sharp lookout for weeds, removing them immediately when they appear. Then thin the plants to stand two inches apart. In about eight weeks they will be ready for transplanting to the garden, where they should be spaced 6 to 10 inches apart in rows 24 to 36 inches apart. To mitigate somewhat the shock of transplanting, try to start the operation on a gray cloudy day. After the plants are set, sprinkle generously with water. If the celery plants are oversized, it is well to pinch off a few of the outer leaves before planting. For all transplants, the most critical period is the first few days. Celery, especially, will have a much greater chance of producing a satisfactory crop if the young plants are protected from the direct rays of the sun for the first day or two. One way of doing this is by inserting small leafy branches into the soil next to the celery stalks. The leaves will provide a mottled shade without reducing the much-needed air circulation.

Care and Feeding Because of its insatiable appetite for nutrients, celery requires a more than adequately prepared soil. Two wheelbarrow loads of a ripe barnyard manure (if available) per 25 feet of row are hardly an overapplication. If manure is scarce, invade the compost pile and take plenty of well-composted material to work into the soil. In addition, about five pounds of a balanced commercial fertilizer, (5–10–10) should be thoroughly worked into the soil. In order to prevent injury to tender root systems, try to complete these soil manipulations at least two weeks before planting. Nutrients can be kept in good supply by applying side dressings of nitrate of soda two or three times at three-week intervals. About one-fourth pound per 25 feet of row should make the difference between a good crop and a mediocre one. Be sure to keep the BB-sized white granules at least four to eight inches away from the row and well mixed in with the soil. Because the root system is not broad or deep, more water is required than for other vegetables. Keep the soil always moist.

Harvesting Although the practice is dying out, some commercial growers continue to blanch celery, a process which produces a crunchy, near-white product. This is done by shielding the celery plants from the sun by the use of upright boards or other material. Many home vegetable growers aren't interested in blanching because they believe that the green chlorophyl in sun-drenched celery is necessary to the vitamin content. It takes very little effort to blanch celery, however, and if the family enjoys crisp white stalks, here is how to get them: Place 12-inch wide boards upright on each side of the row of celery and secure them with short stakes driven into the ground on the side away from the celery (the celery supports the other side). In about ten days the first crop of celery can be taken to the dinner table.

Insects and Diseases Celery may be attacked by a number of fungi, including pink rot, late blight, damping-off, and yellow blight. As a precaution, avoid successive plantings of celery, lettuce, or cabbage in the same soil, stay away from celery plants when they are wet, and sow seed in a disease-free material such as sphagnum moss. Another common pest is the celery leaf tier, an insect that encases itself in leaves rolled and stitched together with webs. Two doses of pyrethrum dust, one to drive the insect from the web and one to dispatch the leaf tier itself, will usually work. For other problems and their remedies, see the chart in chapter 5.

CHINESE CABBAGE (Celery Cabbage)

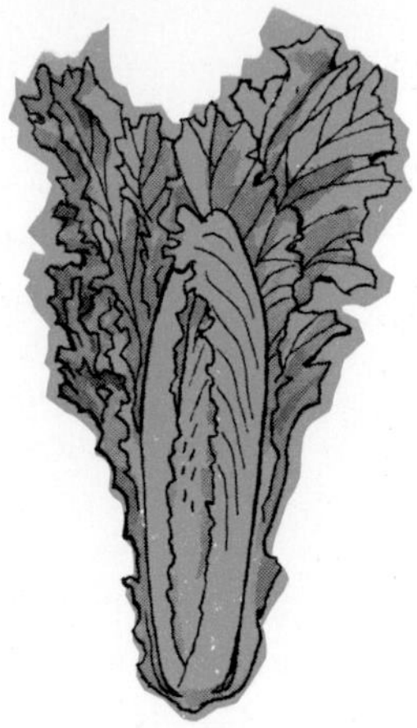

Varieties
Michihli

HARDINESS: *Hardy*
TIME TO PLANT: *Midsummer*
SPACING: *rows 18 to 30 inches*
plants 6 to 8 inches
DEPTH TO PLANT SEED: *½ inch*
SEEDS/PLANTS PER 25′: *1 packet/30 plants*
AMOUNT PER PERSON: *10 feet*
YIELD PER 25′: *25 pounds*
TIME TO HARVEST: *50 to 80 days*

Chinese cabbage has suffered its own identity crisis. Often referred to as celery cabbage because it looks like cabbage and tastes like celery, it is really a member of the mustard family. Strictly a cool-season vegetable, Chinese cabbage should be grown as a fall crop. Warm weather will only slow development and encourage seed stalks to "bolt" (go to seed) rather than produce full leafy heads.

Planting The cultural requirements for Chinese cabbage are similar to those for cabbage. Both need a deep rich soil containing abundant organic matter and nitrogen with plenty of moisture during growth. Sow the seeds in midsummer in a spot vacated by an early-season crop. As plants are thinned to stand six to eight inches apart, use the removed plants for interesting flavor additions to salads. If restricted space has you stymied, exploit the crinkly texture of Chinese cabbage for bordering a flower bed or breaking up the monotony of a stand of ornamentals.

Care and Feeding Because Chinese cabbage is a leaf crop, the soil should get extra amounts of nitrogen-rich organic or chemical fertilizer. When plants are six to eight inches high, sprinkle nitrate of soda along both sides of the row and rake into the soil. For weed control apply a weed-free open mulch, three inches deep. Water frequently to keep the soil's moisture content high.

Harvesting Resembling Swiss chard in appearance, Chinese cabbage forms small heads and wide outer leaves that narrow into much lighter-colored stems as they approach the root portion. As with any leaf crop, the heads should be picked before the leaves become old and tough. Remove the head by slicing the stem at the soil line with a kitchen knife.

Insects and Diseases The young leaves are often attacked by flea beetles and leafhoppers. Both can be effectively controlled with an application or two of relatively nontoxic rotenone. Aphids may also appear. See the chart in chapter 5 for controls.

CORN

Varieties
Royal Crest
Spring Gold
Early Sunglow
Sprite
Butter and Sugar
Gold Cup
Golden Jubilee
Seneca Chief
Silver Queen (white)
Golden Bantam
and many others

HARDINESS: *Tender*
TIME TO PLANT: *After danger of frost*
SPACING: *rows 34 to 42 inches*
plants 8 to 12 inches
DEPTH TO PLANT SEED: *½ inch*
SEEDS PER 100′: *1 packet*
AMOUNT PER PERSON: *30 feet*
YIELD PER 100′: *60 ears*
TIME TO HARVEST: *63 to 100 days*

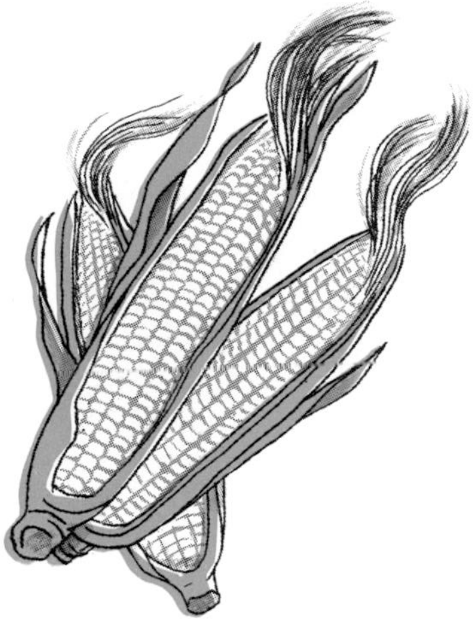

Unless you can afford to squander space, the corn patch will have to be kept on the small side. But many gardeners believe that any amount of garden juggling or just plain lawn grabbing is worth the indescribable taste of your own fresh corn rushed from stalk to boiling water.

Planting Full sun, warm nights, and an abundance of moisture are essential. First get your soil into good condition by working in generous amounts of rotted manure or compost. Then, just before planting the seed, rake a commercial fertilizer (5–10–5) into the topsoil or place in a band three to four inches below the furrow. Be certain that fertilizer granules do not come into direct contact with the seed or injury will result. Corn likes a soil pH of 5.6 to 6.8—slightly acid.

For ease in cultivation and for proper pollination, the ideal placement of corn is in rows rather than hills. Whether your garden area is square, oblong, or pie shaped, remember that corn pollinates better when planted in numerous short rows rather than in one long row. Plant seeds for the first crop about four inches apart, one-half to one inch deep, in rows spaced anywhere from 34 to 42 inches apart, a week before the probable date of the last killing frost (see map on page 36). Plan on succession planting; a new row planted every 10 or 14 days until midsummer will guarantee a steady supply. Or plant an early, a midsummer, and a late variety all at the same time. Early Sunglow, for example, matures in approximately 63 days, Butter and Sugar in about 78 days, and Golden Bantam in 85 days. When sprouts are six to eight inches high, thin out the weaker ones, leaving 8 to 12 inches between plants.

Care and Feeding Once sprouted, corn practically explodes out of the ground, which means huge amounts of nutrients are being drained from the surrounding soil. Try to keep up with the demand by applying a 5–10–5 fertilizer or nitrate of soda (two pounds of complete fertilizer, one pound of nitrate of soda per 100 feet of row) when stalks are 6 to 12 inches high. Scatter fertilizer on both sides of the row and rake into the soil immediately. Be gentle with nitrate of soda—overdoses can injure roots and damage the young plants. Since weeds also compete for nutrients, they should be removed as soon as they appear. As you hoe, try to gather soil up against the stalks to provide added support.

Harvesting Corn ears should not be allowed to ripen beyond the tender milky stage. With a little practice, ripeness can be determined by gently pressing the sides of the ear. Corn ready for the pot will feel firm and full. At the same time the silk at the tip of the ear will be dry and rusty to almost black in color (depending on variety and climate). To remove the ear, simply grip near the base and with a twisting, downward motion break it away from the stalk. Remember, the longer corn is kept before cooking, the tougher it gets. Within 30 seconds after pulling, the sugars begin to turn to starch. Rule of thumb: start the pot boiling *before* you pick the corn, or at least as you begin to shuck it.

To conserve soil nutrients, remove stalks that have been stripped of ears. Bend the stalk over with one hand and with the other cut the stem four to six inches from the ground with a healthy stroke of a hand sickle or machete. Once the harvest is completed, all stalks should be cut and removed. Refuse should be either disposed of or put through a shredding machine for addition to the compost pile.

Insects and Diseases Still major nuisances are the European corn borer and the earworm. The borer is pink or pale brown in color with a dark brown head. About one inch in length, it will feed in stalks and later go down into ears. Look for small holes in the base or side of ears as evidence of borers' presence. Sometimes they will attack tassels and can be detected easily by a broken or bent stalk. The best control method is to eliminate them by hand. If carbaryl or Diazinon spray is used, begin treatment when the first corn is 18 inches high and repeat at least three times at five-day intervals. For other pests and diseases, see the chart in chapter 5.

CUCUMBERS

Varieties
SLICING
Ashley
Marketmore
PICKLING
SMR 18
SMR 48

HARDINESS: *Tender*
TIME TO PLANT: *After danger of frost*
SPACING: *hills 6 to 8 feet*
plants 3 to 5 per hill
DEPTH TO PLANT SEED: *½ inch*
SEEDS/PLANTS PER 25′: *1 packet/12 plants*
AMOUNT PER PERSON: *10 to 15 feet*
YIELD PER 25′: *25 pounds*
TIME TO HARVEST: *55 to 65 days*

Although cucumbers need room to ramble, space should be made in every garden for at least one cucumber vine. But don't despair if the crunch of civilization dictates nothing more than a minigarden. Cucumbers will grow admirably in a discarded nail keg or water barrel, providing good soil is supplied along with adequate drainage. Or they may be trained on fences and walls.

Planting Rich mellow soil that supplies a steady source of moisture is a primary requirement for cucumbers. Plant a cover crop, and then in the spring cover with manure or compost, and a 5–10–5 fertilizer before plowing under. At planting time, additional amounts of old manure or compost (up to four wheelbarrow loads per 50 feet of row) and fertilizer (two pounds per 50 feet of row) can be added to the row. Work manure into six or eight inches of soil under each hill or, when planting cucumbers in rows, dig an extra-deep furrow, spread manure or a band of commercial fertilizer, and cover with a thin layer of regular soil before placing seeds. When sowing in hills, which should be at least six feet apart, again put manure or fertilizer under a thin soil layer before planting. Six to ten seeds should be planted in each hill and later thinned to three or four strong plants. Row planting requires the same distance between rows as between hills. But plants should be thinned until at least one or two feet separate them. If the soil looks as if it might bake and block the emergence of tender seedlings, cover the seeds with a mixture of sand and peat moss. Cucumber seed, however, is especially susceptible to injury from contact with commercial fertilizers, so extra care must be taken to insure thorough soil preparation.

Occasionally the rude finger of winter will poke into late spring and threaten cucumber plants with frost. Cones of newspaper held in place with stones or soil will ward off the cold. Straw, cardboard containers, burlap, or especially designed plastic caps available at garden centers will also suffice. At the same time these protectors will offer some relief from the direct, piercing rays of the sun. Every morning be sure to provide ventilation by lifting each covering just enough to allow a free flow of air throughout the day. When plants are established, remove protectors, but be sure there is no danger of frost before doing so.

If space is a problem, simply fill a discarded nail keg or rain barrel with soil mixed with generous amounts of rotted manure or other humus and one-half pound of 5–10–5 fertilizer. Sow five to ten seeds one-half inch deep, and then give the soil a good soaking. When plants are established, thin out. If ample amounts of water are provided at regular intervals, this raised cucumber bed should support four or five plants. Place the barrel in a sunny location. Because the vines are kept off the ground, the fruits will be virtually unblemished. Similarly they thrive when planted against a sun-splashed fence or wall. But whatever the setting, the soil must be supplied with plenty of nutrients as well as moisture.

Care and Feeding A cucumber is composed of 96 percent water, which gives you some clue to its moisture needs. If one section of the garden seems to hold moisture better than another, sow cucumbers there rather than beans, which require good drainage. The watering can should be kept within reach in case of an unexpected dry spell. Cucumbers are an in-between crop, responding unfavorably to cold as well as to extreme heat. In some areas the crop must be reserved for spring and fall planting only. For this, inquire locally to be sure.

Cucumber is one of those plants that will easily become all chin and no forehead if too much nitrogenous fertilizer is applied. To prevent the vines from running amuck and producing little fruit, keep fertilizer applications during the growing season at a minimum. If applications were somewhat heavy at the outset, pinch off the ends of the vines in order to force greater fruit production.

Weeds, of course, will practically devour the defenseless plants. Here mulch is an excellent practice. Weeds are smothered, moisture content in the soil is preserved, and fruits, once they appear, will be kept insulated from the soil. Any "clean" material will do (see chapter 4 for a list of recommended materials and methods of application), providing the thickness of the mulch does not exceed two inches after it has settled and providing it is airy, not compact.

Harvesting One of the delights of cucumbers is that they can be picked at any stage of their development. Preferably the fruit should be picked on the small side, before the seeds inside have had a chance to become solid. Small cucumbers are excellent for pickling, while larger sizes can be used for salads. In any case, the crop should be picked clean every two or three days even if the quantity threatens to inundate the kitchen. If fruits are allowed to remain on the vine, the plant is likely to stop producing altogether. When picking it's best to leave a small portion of the stem on the fruit. Yellow-green oversized cucumbers are a sure sign you have waited too long before harvesting. For pickling: small ones may be used for sweet pickles, larger for bread-and-butter pickles, largest for dill pickles.

Insects and Diseases The spotted and striped cucumber beetles can reduce the cucumber row to rubble in a day if allowed to roam at will among the plants. Only experience will tell you if these pests are a threat in your area. Spraying with a potent pesticide will help (see chapter 5), but a better method is to cover the young cucumber plants with a protective piece of cheesecloth or tobacco netting held down with stones or soil. Once the cucumber beetle has been effectively excluded, the threat of bacterial wilt will also be considerably lessened, since the beetle is notorious for spreading disease. Select and plant varieties bred to be resistant to mosaic and scab diseases. See also the chart in chapter 5.

EGGPLANT

Varieties
Black Beauty
Early Beauty Hybrid
Jersey King Hybrid
Mission Bell

HARDINESS: *Very tender*
TIME TO PLANT: *After danger of frost*
SPACING: *rows 36 to 42 inches*
plants 24 to 36 inches
DEPTH TO PLANT SEED: *½ inch*
SEEDS/PLANTS PER 25′: *1 packet/7 plants*
(packets average 30 seeds)
AMOUNT PER PERSON: *5 plants*
YIELD PER 25′: *50 pounds*
TIME TO HARVEST: *80 to 85 days from transplanting into garden*

Although short on nutrients, eggplant has a flavor unrivaled by anything else in the vegetable patch. In terms of space-saving value the feltlike leaves and lavender blossoms of the eggplant make it a natural for landscaping purposes. Sun is essential. Because of its slow growth—eggplant needs a long growing season of anywhere from 100 to 140 days from sowing—indoor starting is indicated for northern short-season areas.

Planting Like tomatoes and peppers, eggplant should be started in flats indoors at least eight weeks before you plan to set them out in the garden. Plant seeds one-half inch deep in vermiculite or sphagnum moss that has been spread over at least two inches of ordinary garden soil. Water thoroughly and cover the containers. For procedures on indoor planting see chapter 3. Always grow more plants than needed for the garden in order to make up for later losses. If time and space shortages prohibit indoor planting, eggplant can usually be purchased at a garden supply center. When purchasing started plants, check the stems carefully—plants with hard woody stems seldom produce good yields.

When planting outdoors be sure the soil is really warm. Sensitive to the conditions under which it is grown, eggplant will easily languish if not cared for properly. A friable soil chock-full of nutrients will produce full healthy bushes that may yield up to eight or ten fruits each. Apply several wheelbarrow loads of manure or compost to a 50-foot row, plus about five pounds of a complete fertilizer (5–10–5) or an organic substitute like blood meal, fish scraps, or activated sludge.

Choose a gray day for transplanting. Remember that roots can dry out and become injured if left to the elements for too long a time. Dig shallow holes about 30 inches apart in a straight row and fill with water. When the water in the holes has drained, set plants and fill around them. Firm soil around roots, leaving a slight depression to catch water. Cutworms, a constant threat, are nocturnal insects that can flatten a row of vegetables in a single night. To protect the tender eggplant, simply set a collar of stiff paper or aluminum foil around the stem (see page 44). If the sun is harsh and the newly planted crop seems somewhat peaked, cover the plants with protectors for a few days. Don't make these protectors so efficient that they prevent ventilation.

Care and Feeding If the soil has been adequately prepared at the outset, eggplant will require very little in the way of additional fertilizer. In a matter of weeks the plant will develop several branches. Since it is a low-growing, bushy plant that grows laterally instead of vertically, it requires no support. But adequate moisture and weed control are musts. If the sight of a hoe sends shivers up your spine, try a light mulch of salt hay, pine needles, or peat moss. Black sheet plastic is especially useful in northern areas because it helps keep the soil warm. Eggplant is one of the more sensitive plants and reacts quickly to adverse conditions. The best "fertilizer" is a sensitive eye that can detect a thirsty or nutrient-starved plant before permanent damage is done.

Harvesting More often than not, the ambitious eggplant will set more blossoms than it can nourish into large meaty fruits. The result will be quantity instead of quality produce. After the first blossoms have set and the fruits have begun to form, pinch off later blossoms to conserve plant energy. Snip off fruits when about four inches in diameter and very glossy, before their sheen begins to fade. A dull-looking eggplant is dull to the palate too.

Insects and Diseases Occasionally the eggplant lacebug will feed in groups on the undersides of leaves, causing them to turn yellow and brown. If bugs are unchecked the plant will eventually die. A malathion spray will eliminate the pest if care is taken to cover the undersides of leaves. Malathion should not, however, be applied if fruit is to be harvested in less than three days following the spraying. Colorado potato beetles, flea beetles, and hornworms also attack eggplant. For specific remedies see the chart in chapter 5.

LEEKS

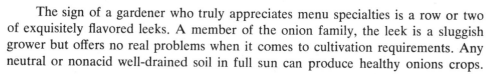

Varieties
Broad London
Large Flag (American)

HARDINESS: *Hardy*
TIME TO PLANT: *When soil can be worked*
SPACING: *rows 24 to 30 inches*
plants 4 to 6 inches
DEPTH TO PLANT SEED: *½ inch*
SEEDS PER 25′: *1 packet*
AMOUNT PER PERSON: *10 to 20 feet*
YIELD PER 25′: *50 plants*
TIME TO HARVEST: *130 days*

The sign of a gardener who truly appreciates menu specialties is a row or two of exquisitely flavored leeks. A member of the onion family, the leek is a sluggish grower but offers no real problems when it comes to cultivation requirements. Any neutral or nonacid well-drained soil in full sun can produce healthy onions crops.

Planting If a crop is to be harvested before the summer begins to wane, leeks should be started indoors for later transplanting to the garden. But leeks that have not reached full maturity will withstand winter's freezing in cold areas and may be taken up in the spring for a presummer treat. Seeds may also be sown in a cold frame where protection from wind and frost will provide suitable conditions for germination and early growth. Simply plant seeds one-half inch deep in rows six inches apart. Once the plants have reached a height of six to eight inches, cut off half the green leafy part and transplant them to the garden (see next page). The same transplanting procedure may be followed when planting seed directly in the ground. It's best to set transplants in a depressed furrow about five to six inches below the normal soil line. Then, as the plants develop, draw soil gradually into the furrow to fill it. Called "blanching," this garden practice will encourage the plants to produce long white tender stems.

Start leeks from seed or put started plants in a furrow 6 to 8 inches deep, filling with soil as they grow. When young plants have reached a height of 8 inches, snip off half of leaf portion and transplant to another row. Blanch stalks by pulling soil up against the plants.

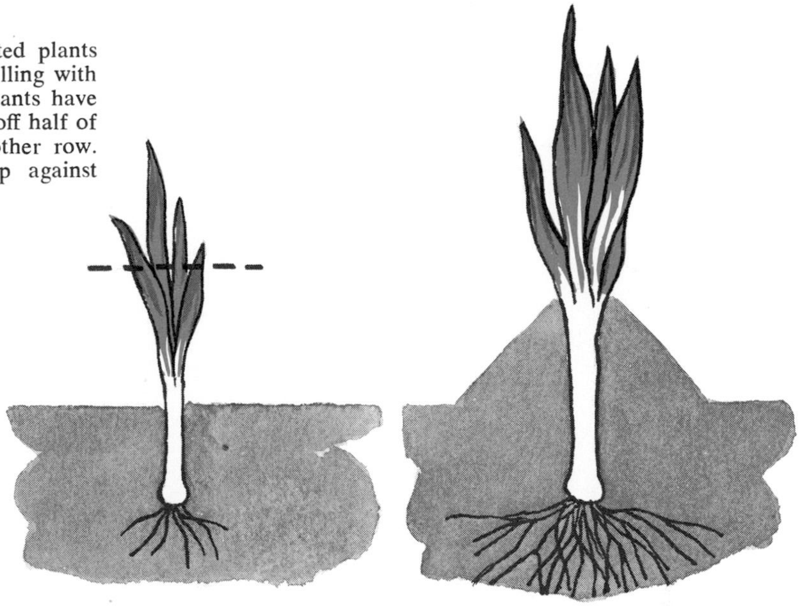

Care and Feeding Leeks, like onions, will respond well to applications of manure or compost. Work materials into the soil along with a scattering of 5–10–5 fertilizer before planting seed. Rake smooth, removing stones and other debris.

Moisture and an abundant supply of plant nutrients are essential if a bountiful crop is expected. The organic way calls for blood meal or fish scraps (nitrogen), phosphate rock or bone meal (for phosphorus), and wood ashes or granite dust (for potassium). Moisture should be supplied at the rate of about one to one-and-a-half inches per week in dry areas or if the weather becomes uncooperative. Weeds will practically blot out the young leek seedlings if not controlled. Since earth should be kept banked against the maturing plants, a mulch will only confuse matters. Use a hoe for shallow cultivating or pull out weeds by hand.

Harvesting When ready for pulling, a ripe leek looks something like a wild and woolly scallion. There is no real peak time. The entire leek plant may be pulled and used at any time. Be sure to replace disturbed soil and bank it up again. Since anything that can be squeezed from the garden in early spring is an extra delight, it is well to leave a few leeks for wintering over.

Insects and Diseases Similar to onions in appearance, the leek is also similar in its susceptibility to onion thrips (see the chart in chapter 5). White blotches on leaves or withering leaf tips are a good sign that thrips are sucking away plant juices. A dusting or spraying of malathion will discourage them. Leeks should not be used until after at least three days have elapsed from the time of the spraying with malathion. For other diseases, see the chart in chapter 5.

LETTUCE

Varieties
HEAD
Bibb
Boston
Buttercrunch
ENDIVE
Deep Heart
Green Curled
ICEBERG
Great Lakes
LOOSE-LEAF
Oak Leaf
Salad Bowl
COS, OR ROMAINE

HARDINESS: *Hardy*
TIME TO PLANT: *When ground can be worked*
SPACING: *rows 12 to 18 inches*
plants 8 to 12 inches
DEPTH TO PLANT SEED: *½ inch*
SEEDS PER 25′: *1 packet*
AMOUNT PER PERSON: *5 feet per planting*
YIELD PER 25′: *19 pounds*
TIME TO HARVEST: *50 to 80 days*

There is no more pleasing sight in summer than freshly sliced tomatoes couched in a bed of crisp lettuce. Easy to grow and hardy to low temperatures, lettuce will thrive in almost any location providing it is protected from excess heat. High midsummer temperatures in most locations will cause the crop to "bolt"—set seed—before decent heads have formed unless a heat-resistant variety is grown. The most rewarding crops come from spring or fall plantings in hot areas. In warmer climates, lettuce can be grown through the winter. An attractive characteristic of lettuce is the refreshing differences between types: head lettuce, which forms into tight "heads"; the curly-leaved or loose-leaf varieties, which produce an abundance of crisp crinkly leaves; and the cos or romaine, which develops thin, tightly folded, upright leaves.

Popular lettuce types are the tall-growing cos, the loose-leaf, and the heading or loose-heading lettuce.

cos lettuce

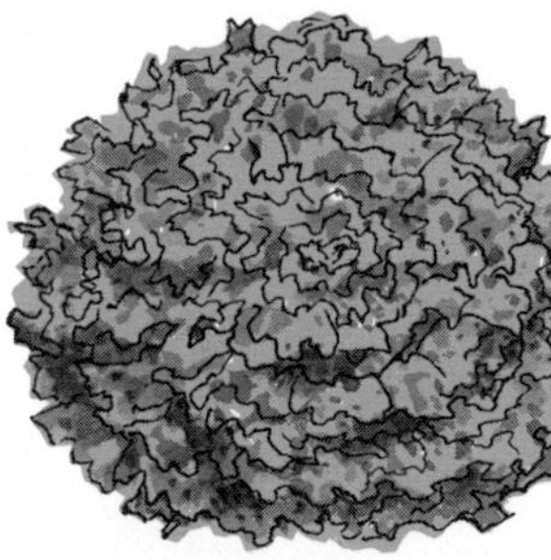

loose-leaf lettuce

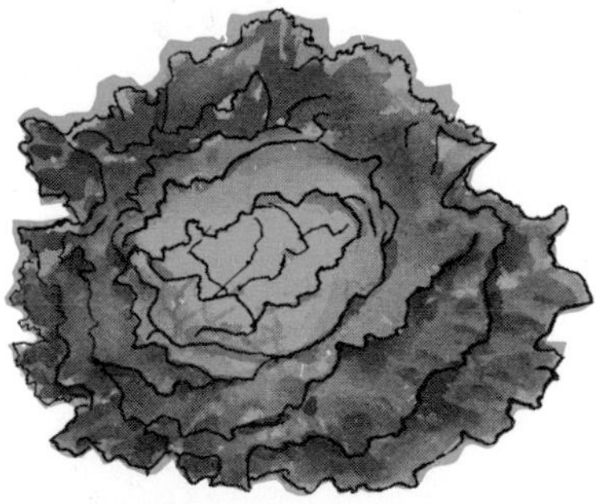

head lettuce

Planting Because of its hardiness, lettuce will undoubtedly be one of the first vegetables to emerge from the garden. An even earlier crop (head lettuce particularly) may be enjoyed by sowing the seed indoors six weeks before the danger of frost usually leaves your area. It's a simple matter to drop seeds into a flat, flower pots, or individual containers. See chapter 3 for indoor planting procedures. Lettuce needs a hardening-off period in a cold frame before final planting in the garden.

When planted outdoors, lettuce is an excellent candidate for intercropping—the sowing of quick-maturing vegetables within the same row or between rows of a slow-maturing vegetable. See chapter 1 for intercropping tips.

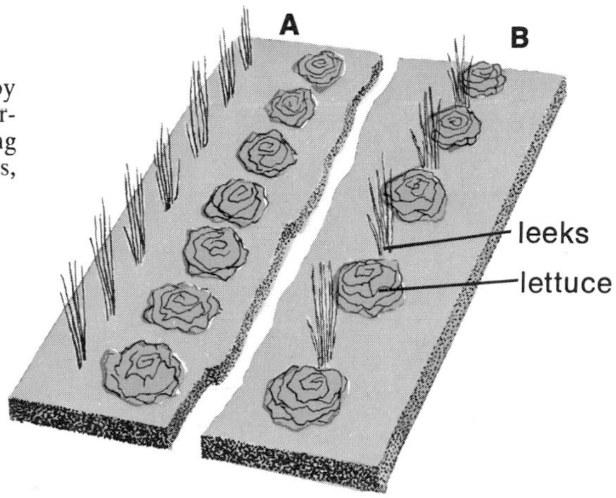

An excellent way to conserve space is by planting lettuce between the rows, or interplanting within the rows, of a slow-growing vegetable such as cabbage, eggplant, leeks, or tomatoes.

The row is ready for thinning when plants reach a height of two inches. Leave one-and-a-half to two inches between plants, being careful not to disturb the root systems of neighboring plants. Thinnings can be used in salads. With good weather and abundant moisture the remaining lettuce plants will quickly eat up all available space and once again appear crowded. At this point another thinning is in order. This time, however, the plants may be either transplanted to another row or used for salads. A third thinning will probably be necessary before the desired spacing of 12 inches between heads is achieved. Loose-leaf types, however, can be as close as six to ten inches apart.

Because of its quick growth, the joys of crisp lettuce in salads can easily come and go before the summer has begun if space is not provided for succession plantings. It is far better to plant short rows at regular intervals of ten days to two weeks than it is to load the garden with amounts that could feed a battalion. Five feet of row per person is plenty. For an average family of three, mark out three 15-foot rows and plant seed every ten days. As one row becomes depleted the next row will be nearing maturity. It's best to pause briefly during the dead of summer when excessive heat makes it difficult to grow a good crop. Begin another series of succession sowings for a fall harvest after the summer heat has reached its peak.

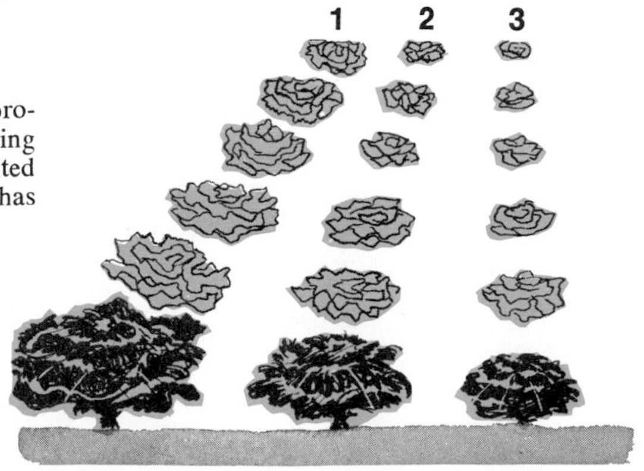

Succession plantings at 10-day intervals provide a continuous crop of quick-maturing lettuce. In practice, row 3 can be planted in space made available after row 1 has been harvested.

Care and Feeding A neutral or nonacid soil is best for most lettuce. Adding a high-phosphorus fertilizer (5–10–5) or what is known as a "superphosphate" fertilizer will contribute to large well-formed heads. A natural source of phosphorus, phosphate rock, will accomplish the same end when spread at the rate of approximately one pound per 100 square feet of garden. And since nitrogen fertilizers cause bursts of leaf growth and lettuce is all leaves, it is a worthwhile practice to scratch in a handful of nitrate of soda or a pound of cottonseed meal along the row now and then. The leaves are the crop, so be certain that fertilizing material is not allowed to remain in contact with the plants. Keep material three to six inches away from the row and work into the soil with a hoe or steel rake. If fertilizer is inadvertently dropped on plant leaves, wash them off well with a sprinkling of water as soon as possible. Root systems of lettuce are small, so both humus and fertilizer, plus plenty of water, are needed.

If endive is grown, and blanched centers are desired, draw the outer leaves of nearly full-grown plants over the center and tie.

Harvesting Although the leaves of lettuce may appear to be crisper in the early morning, it has been proved that the vitamin and nutrient value of lettuce is considerably higher when it is picked later in the day. Pick lettuce when needed, but not until after 10:00 A.M. at the earliest. It is also advisable to pick lettuce on a regular schedule to prevent loss of nutrients through tissue breakdown. Other harvesting practices will be dictated by the variety of lettuce. Head lettuce may not "head" in the hottest weather, but may be harvested even so. When harvesting a heading type of lettuce cut the entire plant where the stem enters the soil. Loose-leaf varieties are harvested by picking the outer leaves, and allowing new leaves to continue growing from the center of the plant.

Insects and Diseases The most troublesome insects to attack the lettuce plant are the cabbage looper and the aphid. The cabbage looper is best controlled by hand-picking. The aphid can be effectively discouraged by spraying the plant with a strong stream of water from the hose. Since the leaves of the plant are what you eat, sprays or dusts should be used with the greatest caution. See the chart in chapter 5 for tips on using sprays.

MUSKMELON (Cantaloupe)

Varieties
Burpee Hybrid
Delicious 51
Iroquois
Samson

HARDINESS: *Tender*
TIME TO PLANT: *After danger of frost*
SPACING: *hills 5 to 7 feet*
plants 3 to 5 per hill
DEPTH TO PLANT SEED: *¾ inch*
SEEDS PER 25′: *1 packet*
AMOUNT PER PERSON: *3 to 5 hills*
YIELD PER 25′: *35 pounds*
TIME TO HARVEST: *85 to 90 days*

This very tender crop needs a long growing season, a light but rich soil, and plenty of sun and moisture. Muskmelon requires a little extra effort, but chilled melon served with a slice of lime is a treat not to be dismissed simply because of an abbreviated season, or the extra attention it needs. If cool short seasons dampen your muskmelon plans, look for early varieties such as Delicious 51.

Planting The cultivation requirements for muskmelon are similar to those for cucumbers, which is one reason why the melon is classified as a vegetable though it is eaten as fruit. Preferring a light-textured soil with good drainage, muskmelon should not be planted outside until all danger of frost is past. Manure or other organic material added to the soil will contribute to the formation of large, well-formed fruit. Before planting, place several shovelfuls of manure or compost under each hill. The most convenient planting method consists of sowing six to eight seeds in hills spaced about six feet apart. If the soil tends to bake and form a hard crust, mix a small quantity of peat moss or leaf mold into the top layer of soil. Thin the seedlings to three to five plants per hill. If a cold snap threatens, protect with a covering of plastic, newspaper, or inverted bushel baskets, weighted down to prevent wind from displacing them. In short-season areas, plant early-maturing varieties or start seeds indoors in pots three to four weeks ahead of outside planting time. See chapter 3 for tips on indoor planting.

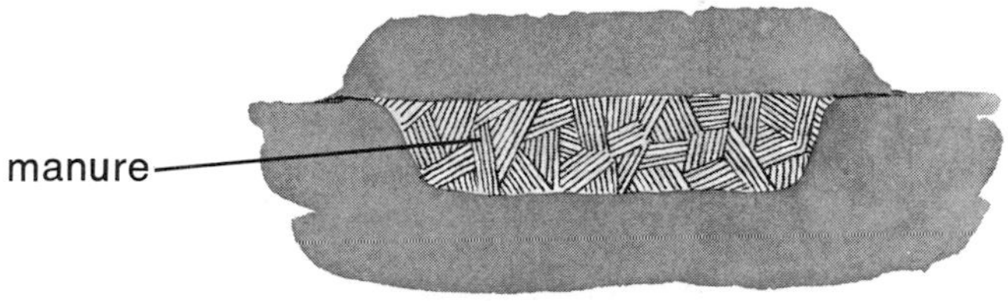

Muskmelon will grow better when compost or rotted manure is placed beneath hill; cucumbers benefit from this treatment too.

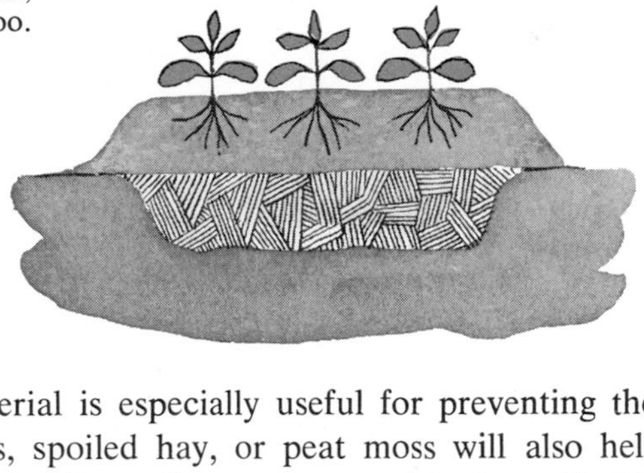

Care and Feeding A mulching material is especially useful for preventing the inevitable weed invasion. Wood chips, spoiled hay, or peat moss will also help to retain soil moisture and protect fruits from direct contact with the soil. Muskmelons, like beets and lettuce, are fairly sensitive to acid soils and will not develop properly in them unless a liming program is undertaken in early spring. See chapter 2 about testing soil pH and corrective measures.

Harvesting Muskmelon might well be the trickiest crop of all to harvest because of the practice it takes to become adept at recognizing a ripe melon. The experienced grower can determine ripeness by the amount of pressure required to separate stem from fruit. Ideally the stem should slip easily from the fruit without resistance. If force is needed the melon is probably not at peak ripeness. Some gardeners know a ripe melon by its aroma.

In cool areas, growth can be accelerated by placing boards, squares of tar paper, or black plastic under developing fruits. But keep a vigilant eye on the melon patch as the plants mature to make sure melons don't become overripe.

Place a board or square of tar paper under each muskmelon to hasten ripening.

Insects and Diseases If the leaves become curled and distorted it is a sure sign that aphids are present. In addition to sucking the vital juices out of the plant, aphids are carriers of virus diseases. When the plants are dry, spray with malathion, being certain to cover the undersides of leaves where aphids collect. Treatments may have to be repeated on a weekly basis before the pest is eliminated. Cucumber beetle and other insects and diseases are listed, along with symptoms and recommended remedies, in the chart in chapter 5.

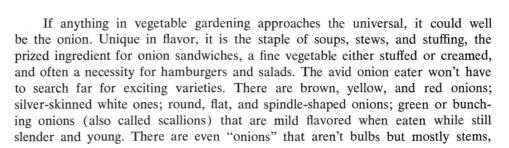

ONIONS

Varieties

NORTHERN

Bunching Onions, or Scallions
Early Yellow
Early Yellow Globe
Sweet Spanish (plants)
Yellow Globe Danvers

SOUTHERN

Crystal White Wax
Excel
Texas Grano

HARDINESS: *Very hardy*
TIME TO PLANT: *When soil can be worked*
SPACING: *rows 12 to 24 inches*
plants 3 to 4 inches
DEPTH TO PLANT SEED: *¾ inch*
SEEDS/PLANTS PER 25′: *1 packet/100 plants or ½ pound of sets*
AMOUNT PER PERSON: *10 to 20 feet*
YIELD PER 25′: *25 pounds*
TIME TO HARVEST: *100 to 130 days*

If anything in vegetable gardening approaches the universal, it could well be the onion. Unique in flavor, it is the staple of soups, stews, and stuffing, the prized ingredient for onion sandwiches, a fine vegetable either stuffed or creamed, and often a necessity for hamburgers and salads. The avid onion eater won't have to search far for exciting varieties. There are brown, yellow, and red onions; silver-skinned white ones; round, flat, and spindle-shaped onions; green or bunching onions (also called scallions) that are mild flavored when eaten while still slender and young. There are even "onions" that aren't bulbs but mostly stems, such as leeks and chives. Some are very mild, others pungent. In short, there is an onion for every palate and every garden.

Highly resistant to frost, onions will grow abundantly under a wide variety of climatic conditions and in almost any section of the country. In terms of the space-yield ratio, onions are perfect for small gardens because they require little space, produce well, and can be stored. Soil conditions, however, are crucial. Moisture and neutral or nonacid soil, both friable in structure and free of stones, are necessary if well-formed onion bulbs are expected.

Planting There are three ways you can get your onion patch started: by planting sets or seedling plants, or by starting from seed. Onions sets are small bulbs sold by the pound or raised from late seed sown the year before and saved. They produce earlier onion crops than those sown in spring from seed. If purchased, make sure the bulbs are not over three-fourths inch in diameter.

A

B

C

Onions can be planted in following ways: (A) from purchased onion sets or sets preserved from the previous season, (B) from seedlings started from seed in a cold frame for later transplanting, (C) from outdoor-sown seed (this method requires a long growing season).

PLANTING FROM SETS—Make rows 12 inches apart for hand cultivation, 24 inches apart if a machine cultivator is to be used. With the hoe, dig a shallow trench one inch deep, using taut twine as a guide. Plant the sets four inches apart, making sure the pointed end is facing up so leaves can sprout directly into daylight and roots will be within easy reach of moisture. Replace the soil and tamp gently to prevent air pockets from forming around roots. For delicious green onions, plant sets two inches apart; then during the growing season pull every other plant for salads. Any good seedsman or garden supply center will have plenty of sets on hand when spring weather approaches.

PLANTING SEEDLINGS—These small plants, about the size of scallions, are handled in the same way as onion sets. They are available locally or by mail from seedsmen. Plant seedlings four inches apart in shallow furrows two inches deep as soon as the soil can be worked.

PLANTING FROM SEED—In northern sections where summers are cool, onions can be planted from seed with satisfactory results as soon as the ground can be worked. An obvious advantage of seeds is their low cost compared to either sets or seedlings. But the crop is often uneven and not as high in quality. Thick sowings can improve the harvest as well as provide intermediate pickings in the form of thinned-out plants.

Whatever method is decided upon, onions need a well-prepared soil, one rich in phosphorus and potash. If your soil seems compact and is difficult to manage, consider fall cultivation to allow conditioning of the soil through alternate freezing and thawing. In spring add generous amounts of manure or compost and work into the topsoil with a spading fork or machine cultivator. Just before planting time add a commercial fertilizer high in phosphorus and potassium (5–10–10) or organic substances high in these nutrients. If soil is acid, liming is needed. See chapter 2 for appropriate materials and rates of application.

Care and Feeding Because they do not develop broad heavy leaves that cast a dense shade, onions are easy prey for invading weeds. Frequent hoeing plus hand-picking of weeds growing within the row will conserve nutrients. When the plants have become established and are from six to eight inches high, apply a side dressing of a nitrate fertilizer. Work granules into the soil with a hoe or rake. A gentle sprinkling of water will dissolve the fertilizer and rush valuable nutrients to plant roots.

Harvesting When the rows are densely planted, a scattered crop of onions can be enjoyed as rows are thinned during the early stages of the growing season. Otherwise the onions must be left to mature at their own practically invisible rate. When the tops bend and flop over in late summer, the crop is nearing the harvest stage. The tops will turn brown and become pinched where they connect with the bulb. This is the time to pull the onions and spread them out in a garage or shed to cure for a few days, before removing the tops and storing the bulbs in a cool dry place. Keep them well ventilated by storing in shallow open trays or in mesh bags suspended from the ceiling.

Most onions keep well in dry storage for a period of time. The varieties that keep well all winter are Early Yellow Globe and Yellow Globe Danvers.

Insects and Diseases The most disagreeable pest to attack the onion is the onion maggot, a small white legless insect that burrows into bulbs and is found mostly in northern portions of the country. An organic control for maggots that may work is to plant radishes on both sides of the row. Maggots will probably be attracted to the quick-growing radishes, which can then be pulled and destroyed. An application of Diazinon will offer some control. Allow at least three days between the time of spraying and harvest. Occasionally onion thrips will suck at the leaves, leaving small white blotches. If the thrips are unchecked, the leaves will eventually wither and turn brown. A dust or spray of Diazinon will effectively discourage thrips. See also the chart in chapter 5.

PARSNIPS

Varieties
Hollow Crown
Model

HARDINESS: *Very hardy*
TIME TO PLANT: *Early spring*
SPACING: *rows 18 to 30 inches*
plants 3 to 4 inches
DEPTH TO PLANT SEED: *½ inch*
SEEDS PER 25′: *1 packet*
AMOUNT PER PERSON: *10 to 15 feet*
YIELD PER 25′: *½ bushel*
TIME TO HARVEST: *120 days*

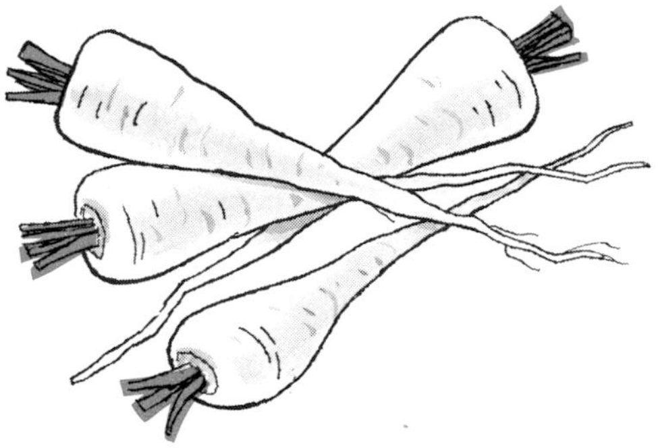

Unlike most vegetables, parsnips actually benefit from prolonged exposure to lower temperatures, making them an ideal cold-area crop. If left in the ground over the winter in the North and Northeast, they improve in taste and provide fresh produce from the garden while stubborn patches of snow still linger in the shade. They may also be stored indoors. Their delicate sweet flavor makes them excellent additions to stews or soups. Or they can be served mashed or sliced and deep-fat fried. In southern and southwestern areas, plant in the fall and grow as a winter crop.

Planting Since the edible root of the parsnip can grow to a depth of 15 inches, it is important that the soil be worked deeply and thoroughly, and raked free of stones and twigs. Sow seed thickly in rows 18 to 30 inches apart. Since germination is slow, it helps to cover the seed with leaf mold, peat moss, or a mixture of sand and soil. Tamp soil firmly over the seed to speed the germination process. When seeds have sprouted and plants have grown to a height of three or four inches, thin to three inches apart.

Care and Feeding Parsnips need rich soil, but if manure is incorporated into the topsoil it should be well broken down and free of lumps. Any kind of obstruction will cause the roots to fork and become distorted in shape, so all debris should be removed from soil. Weeds will have to be removed as soon as they appear. Since germination time is prolonged, it is wise to mix small amounts of rapidly germinating radish seed with the parsnip seed to mark the row among weeds.

Once the parsnips have sprouted, their delicate carrotlike tops are easily distinguishable from the broader-leaved weeds. A mulch can be a valuable ally in the struggle against weeds; it will also protect the parsnip roots if they are left in the ground to winter over.

Harvesting When the tops of the roots have reached a diameter of two to three inches, the crop is ready for harvest. It's best, however, if you live in a cold area, to delay picking until the row has been nipped by at least one sharp frost. If a spring crop is desired, simply allow the roots to remain over the winter, protected by a layer of mulching material. As soon as the sprouting of new growth is detected, picking can be started. New growth should not be permitted to continue for too long. The quality of the crop will be preserved if all plants are dug then and stored in a cool place.

Insects and Diseases Some natural controls suggested are *not* to plant parsnips near carrots, celery, or caraway, since insects and other troubles that affect them also affect parsnips. Also, companion plantings seem to help keep troubles away. Plant them with bush beans, peppers, potatoes, and peas, and also with radishes, onions, and garlic. For specific pests and diseases, consult the chart in chapter 5.

PEAS

Varieties
Freezonian
Frosty
Progress No. 9
Wando (hot weather resistant)

HARDINESS: *Hardy*
TIME TO PLANT: *Early spring*
SPACING: *rows 24 to 36 inches*
2 to 3 inches
DEPTH TO PLANT SEED: *1 to 2 inches*
SEEDS PER 25′: *¼ pound*
AMOUNT PER PERSON: *50 to 100 feet*
YIELD PER 25′: *10 pounds*
TIME TO HARVEST: *50 to 68 days*

Productive, easy to grow, and hardy, peas require little more than cool temperatures and a fertile soil. The first step is to study the varieties of peas available. Some are a dwarf bush type that requires no support. Taller-growing varieties need a trellis or brush stuck into the ground for the vines to climb on. The edible-pod varieties produce soft stringless jackets that can be cooked and eaten like string beans when young or, if preferred, the peas can be allowed to mature like other varieties, then shelled and cooked. Others are good for freezing

Planting Since peas will not thrive in the heat of midsummer, they should be planted as early in the spring as soil conditions permit. Where climate allows, plant early and late varieties a week apart. Work the soil thoroughly, adding ample amounts of manure or compost. To grow productive plants, try digging the furrow several inches deeper than the recommended planting depth and fill with compost or a low-nitrogen commercial fertilizer (5–10–10). Then, on the three-to-four-inch-wide flat-bottomed trench, scatter peas for a full bushy stand and cover with an inch of soil. After sprouts appear, thin plants to an inch apart in the row. Pea seeds are susceptible to fertilizer injury, so keep fertilizer well away from seed. Chicken wire fastened to stakes spaced six feet apart along the row or tall brush inserted next to plants will give peas something to cling to as they mature.

Chicken wire fastened to stakes makes an excellent support for climbing pea plants. After harvest, roll up wire and stakes and store for the next season.

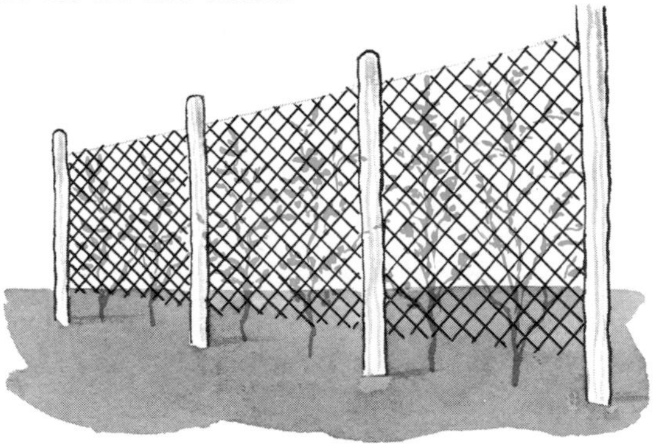

Brush pushed into the ground along the row is a less expensive method and offers sufficient support for tall-growing varieties of peas.

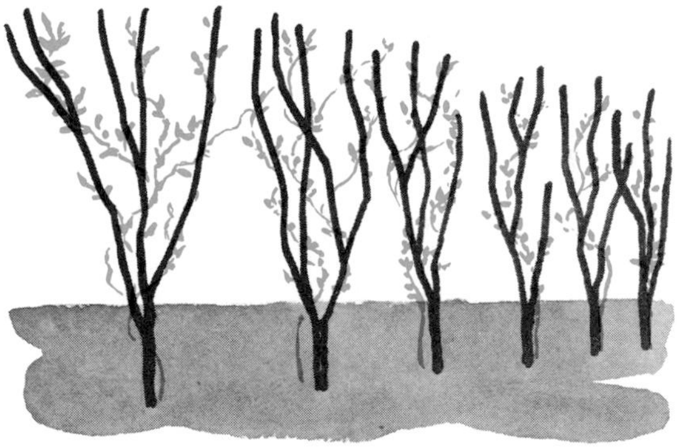

Care and Feeding A side dressing of a complete low-nitrogen fertilizer when peas have reached a height of six to eight inches will give them the necessary boost to mature into heavy-yielding plants. Sprinkle fertilizer sparingly on either side of the row, taking care to keep granules off the leaves. One application is plenty.

Hold weeds to a minimum by frequent hoeing or the application of a mulch. The taller-growing varieties are excellent candidates for mulching, but will, in addition, require some high form of support (see above).

Harvesting The trick to picking peas is to remove the pods just before the peas reach full size, but when the pods are firm and well filled. Gently press the pod between the fingers. If it feels solid it is ready for picking. Since the pods at the bottom of the plant reach maturity first, the pickings should always start at the bottom and work upward. When the last picking has been taken from the top, the plants are ready to be removed to make space for a following crop of cabbage, beets, carrots, or lettuce—even two crops if quick-maturing ones are chosen.

Insects and Diseases Aphids and pea weevils are the common insects that attack peas. The pea weevil, about one-third inch long and brown, white, or black in color, can be removed by hand. Aphids are often discouraged by a blast of water from the garden hose. Malathion is an effective spray against both pests. See the chart in chapter 5 for further controls.

PEPPERS

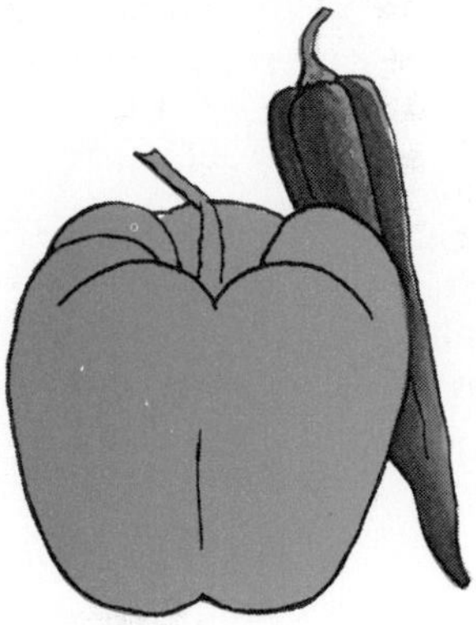

Varieties
Vinedale (early)
Calwonder
Yolo Wonder
Bell Boy Hybrid

HARDINESS: *Very tender*
TIME TO PLANT: *After danger of frost*
SPACING: *rows 20 to 36 inches*
plants 18 to 24 inches
DEPTH TO PLANT SEED: *½ inch*
SEEDS/PLANTS PER 25′: *1 packet/about 12 plants*
AMOUNT PER PERSON: *4 plants*
YIELD PER 25′: *10 pounds*
TIME TO HARVEST: *60 to 80 days*

A decidedly warm-weather vegetable, the pepper joins the list of crops that should not be planted until after danger of frost is past. If space is tight in the garden, peppers do well as border plants in the flower garden and offer interesting foliage variations to the landscaping. The most popular pepper type is the sweet variety used for flavoring, and as a hot vegetable or salad garnish. Hot peppers are not often grown in home gardens except in the Southwest and Far West.

Planting To guarantee a long enough season, either start seed indoors or buy greenhouse-started plants at planting time. Sow seeds in flats indoors about six to eight weeks before outdoor planting time. See chapter 3 for tips on starting plants indoors. The young plants should be given two to three weeks of hardening-off treatment in the cold frame before final planting in the garden. Planting and soil requirements are the same as those for the tomato (see page 121).

Care and Feeding Because peppers are not heavy feeders, additional fertilizing is not necessary. In fact, too much nitrogen will cause overleafing with a resulting reduction in yield. Use a mulch to smother weeds and keep moisture in.

Harvesting Peppers can be used at any stage of development. But it's more economical to wait until they have reached full size. For bright red peppers, leave them on the plant until they mature and turn a uniform red.

Insects and Diseases It is unlikely that an insect or disease will present a serious problem. Occasionally flea beetles, leafhoppers, or potato beetles might invade pepper plants, but they can easily be controlled with a single application of an all-purpose spray. For specific pests and controls, see the chart in chapter 5.

POTATOES

Varieties
EARLY
Irish Cobbler
Norchip
LATE
Katahdin
Kennebec

HARDINESS: *Semihardy*
TIME TO PLANT: *When soil can be worked*
SPACING: *rows 36 to 42 inches*
plants 12 to 15 inches
DEPTH TO PLANT SECTIONS: *4 to 6 inches*
SECTIONS PER 25′: *20*
AMOUNT PER PERSON: *25 feet*
YIELD PER 25′: *30 pounds*
TIME TO HARVEST: *100 days*

Because of their high yield per plant, potatoes are a good investment of garden space and are little trouble to grow. Potatoes also store well, so if space permits put in extra rows for the cellar potato bin. Even small gardens should have a few plants to let the family enjoy the taste of potatoes, especially new potatoes, swimming in butter.

Planting Potatoes will thrive in any well-drained fertile soil, but will do best in a slightly acid soil where a disease called "scab" is not as likely to affect them. Manure is a bonus to potatoes, but it should be either well rotted or worked into the soil at least two weeks before planting. Another good way to supply nutrients is by spreading a band of a 5–10–10 commercial fertilizer along the bottom of the furrow. Use about seven pounds for every 100 feet of row and cover with at least three inches of soil. If your soil is already fertile, reduce the amount to four pounds for every 100 feet of row. Do not plant in soil that has recently been limed or treated with wood ashes.

Some gardeners cut up old winter potatoes that are starting to sprout, but for best results obtain certified seed potatoes from a garden center or farmers' outlet. These are treated to be disease resistant. Cut the potatoes into sections, making sure that each section contains at least one or two "eyes" or sprouts. Plant to a depth of four to six inches and cover with enough soil so that a slight mound remains. Sections can easily rot if the soil is too moist and too cool, so resist the temptation to plant them too early, before the soil is warm and dry.

Planting Sections of Seed Potatoes

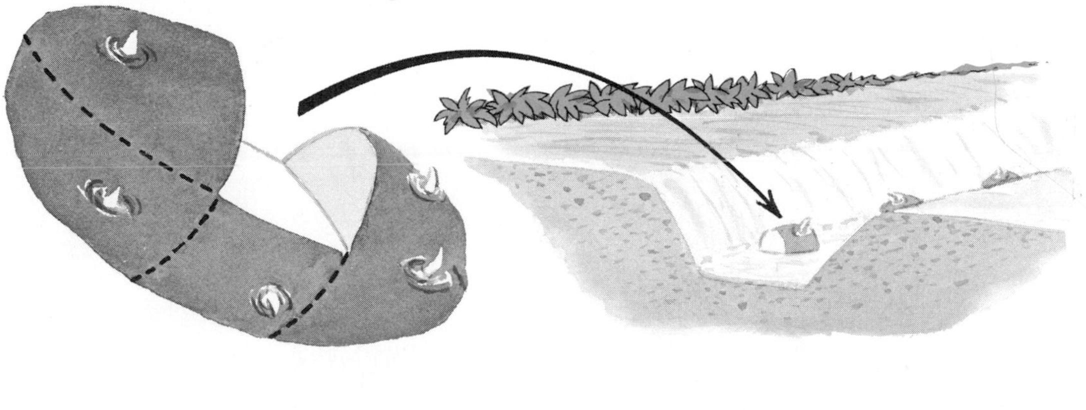

Care and Feeding Once sprouted, the potato vines will grow rapidly on their own. But weeds must be nipped immediately and the soil kept in loose friable condition. Periodically hoe the area between rows, but take care not to injure the roots. Keep the developing tubers covered with plenty of soil to shield them from sunlight, which will turn them green. If necessary, pull soil up around the plants with a hoe or rake.

Harvesting One of the best things about potatoes is that they can be used even when small, although for larger crops they should not be dug until the vines begin to die back. Use a spading fork to dig tubers carefully from the soil. Storage potatoes should receive extra-special handling, since bruises or scratches will shorten their storage life considerably. Harvesting is much simpler when the soil is not wet, so pick a sunny dry day for digging. There will be many small, walnut-sized "new" potatoes near the surface. Set them aside to be enjoyed by the family immediately.

Insects and Diseases The most common pest of the potato is the Colorado potato beetle, the "potato bug" whose grubs can defoliate a plant in a matter of days. And a plant without leaves cannot live long. As soon as the pinkish, bulbish grubs appear on the leaves, spray with an all-purpose spray or use Sevin, endosulfan, or malathion. Hand-picking beetles and crushing egg masses are also effective if done often. See the chart in chapter 5 for other ills and their cures.

PUMPKINS

Varieties
Big Tom
Cinderella
Small Sugar

HARDINESS: *Tender*
TIME TO PLANT: *After danger of frost*
SPACING: *hills 4 to 6 feet*
plants 2 to 3 per hill
DEPTH TO PLANT SEED: *1 inch*
SEEDS PER HILL: *6 to 8 (1 packet)*
AMOUNT PER PERSON: *1 hill*
YIELD PER 25′: *10 pumpkins (or more)*
TIME TO HARVEST: *95 to 115 days*

As a rule, pumpkins need much more room than the average gardener can afford. But they have one thing in their favor, which may enable you to find a place for a vine or two—they will tolerate a moderate amount of shade. Try growing them along a fence or between rows of corn.

Planting Sensitive to extremes of both cold and heat, pumpkins will not thrive in the heat of a southern midsummer or in the cool dampness of a northern spring. When the weather has settled and the soil is fairly warm, plant six to eight seeds per hill in hills four to six feet apart. After seedlings have emerged and have become established, thin out the weaker plants, leaving two or three vines per hill. If you decide to mingle pumpkin with corn, space the plants eight to ten feet apart in every third or fourth row.

Care and Feeding The pumpkin patch is a good place to use a mulching material, because the ranging vines make it difficult to cultivate them with a hoe or machine. In addition to smothering weeds, the mulch will keep pumpkins clean and help preserve soil moisture.

Harvesting Allow pumpkins to become thoroughly matured on the vine, but protect from frost or harvest before frost. Cut off fruit, leaving a portion of the stem attached. To harden skins it helps to leave pumpkins in the field a few days before gathering for storage. Or they may be taken before frost and piled in the garage or on a covered porch to ripen.

Insects and Diseases Pumpkins are relatively safe from serious damage due to insects or disease. They are sometimes bothered by the striped cucumber beetle, the squash bug, or the vine borer, but all can be quickly controlled with an all-purpose spray. See also the chart in chapter 5.

RADISHES

Varieties
Champion
Cherry Belle
Sparkler
White Icicle

HARDINESS: *Hardy*
TIME TO PLANT: *When soil can be worked*
SPACING: *rows 12 inches*
plants 1 to 2 inches
DEPTH TO PLANT SEED: *¾ inch*
SEEDS PER 25′: *1 packet*
AMOUNT PER PERSON: *5 feet*
YIELD PER 25′: *10 pounds*
TIME TO HARVEST: *20 to 25 days*

Radishes reach the eating stage in no time and will be one of the first crops to come from the garden. The secret of having tasty mild radishes is quick growth. If maturation is slow and tedious, they will become hard and pungent.

Planting Radishes are not choosy when it comes to the type of soil they are grown in, but high fertility is essential for rapid growth. Old manure or other fertilizer should be worked into the soil. Since radishes are quick growers, try to plan on a number of succession plantings spaced from seven to ten days apart. If rows are kept small (a six-foot row is enough for an average family) and sowings are continuous, you should be able to enjoy radishes throughout the entire season. See chapter 1 for intercropping and succession planting. Prevent overcrowding of seedlings by planting no more than 15 seeds per foot of row.

Care and Feeding If ample fertilizer is provided at planting time, additional applications will be unnecessary. But quick growth can't occur without sufficient amounts of moisture. Give radishes a dose of water at least once a week—more in dry areas or in droughts—and keep weeds from robbing essential nutrients from the soil. Frequent hoeings will keep the soil loose and in proper condition to receive additional rainfall.

Harvesting Harvesting times vary for each variety. The scarlet and bright red types mature in 20 to 25 days, while the white icicle type that grows like carrots requires 30 to 40 days. The black Spanish radish (actually white-fleshed), which is considered a fall and winter radish, matures in 60 days. But all should be pulled when they have reached a moderate size. Cracking or splitting means radishes have aged too long.

Insects and Diseases The root maggot tunnels into radishes, destroying them for eating. It is most likely present where cabbage or cauliflower is grown, and control is somewhat tricky, requiring an application of six ounces of Diazinon granules for every 100 square feet of garden or along the furrows at least one week before planting. Mix granules thoroughly into the top four to six inches of soil. See also the chart in chapter 5.

RHUBARB (Pie-Plant)

Varieties
MacDonald
Valentine
Victoria

HARDINESS: *Hardy*
TIME TO PLANT: *Spring or fall*
SPACING: *rows 4 feet*
plants 3 to 5 feet
DEPTH TO PLANT ROOTS: *4 to 6 inches*
PLANTS PER 25′: *5 to 7*
AMOUNT PER PERSON: *1 plant*
YIELD: *Varies with location and age*
TIME TO HARVEST: *3 to 4 years*

Rhubarb, grown for its fruit-flavored stalk, does best in regions having cool seasons and winters cold enough to freeze the soil to a depth of two to three inches. One of the easiest and most rewarding of the perennials, it should have a place in every garden where conditions permit. If space is limited, rhubarb will thrive along a fence or wall. A row consisting of five or six healthy plants will provide plenty of tasty stalks for an average family. Though fruitlike, it is classed among the vegetables.

Planting Rhubarb needs a deeply prepared soil supplied with generous amounts of plant food. With a spade or plow, turn the soil to a depth of 12 to 16 inches and work in plenty of manure, compost, or partially rotted mulch material. In spring buy well-established roots and plant them four to six inches deep and three to five feet apart. Cover with soil and sprinkle well with water.

It's also possible, but not common, to start rhubarb from seed and then transplant it. This requires more time—two years to achieve usable stalks—and is not as reliable as planting established roots of a good variety.

Whatever method you decide on, be sure to locate the rhubarb to one side of the garden where it can thrive undisturbed by the yearly digging for annual vegetable plantings.

Care and Feeding Rhubarb is the heaviest feeder of all the garden vegetables, requiring heavy doses of plant food. If manure is difficult for you to come by, commercial fertilizer (10–6–4) will suffice. Scatter about a pound of high-nitrogen food around each hill every year in early spring when it will help force plant growth for an early picking. In the fall use a mulch to cover the roots and keep frost from entering too deeply into the soil.

At some time during the season a seed stem bearing a bulbous seed bud at its tip will appear on each plant. These should be removed immediately. If left to develop, they will divert valuable nutrients from the edible stalks.

In about seven to eight years the plants will become thick and overgrown, and stalks will gradually diminish in size. When this occurs, the plants should be dug up in the fall and the roots split for new plantings. Each root division should have at least one to three buds and a healthy section of root attached. Replant in a different section of the garden. Once planted at four-foot intervals and heavily fertilized, the root fragments will grow into new thick-stemmed plants.

Harvesting Stalks should not be removed until after the plants have developed stable root systems. As a rule a light harvest can be taken the third year—if you are lucky, the second season—and a full harvest by the fourth year. Remove the stem by gripping it near the base and twisting to one side. The stalk should separate easily from the plant below soil level. No more than one-third of the plant should be picked during a season. Detach leaves and add them to your compost heap. Under no circumstances should they be used as food, since they contain injurious substances such as oxalic acid.

Insects and Diseases One of the beauties of rhubarb is that it is relatively free of injurious insects and disease. See the chart in chapter 5 for controls.

SPINACH

Varieties
SPRING
Dark Green Bloomsdale
Hybrid No. 7
FALL
Blight-Resistant Savoy

HARDINESS: *Hardy*
TIME TO PLANT: *When soil can be worked*
SPACING: *rows 14 to 18 inches*
plants 3 to 5 inches
DEPTH TO PLANT SEED: *¾ inch*
SEEDS PER 25′: *½ ounce*
AMOUNT PER PERSON: *20 feet*
YIELD PER 25′: *10 pounds*
TIME TO HARVEST: *48 to 70 days*

For full healthy leaves, spinach should be grown during the cooler portions of the growing season. Resistant to frost, it can often be wintered over if protected by a thick layer of mulch. Like beets, spinach is sensitive to acid soils and if soil tests indicate a pH of below 5.6 to 6.0 (see chapter 2), small amounts of agricultural lime should be worked in regardless of the treatment the rest of the garden has received. The ideal pH for spinach is between 6.0 and 6.5.

Planting When preparing your soil for spinach try to work in about 100 pounds of rotted manure or three to four pounds of a complete high-nitrogen commercial fertilizer (10–6–4) for every 100 square feet of garden area. (A bushel of rotted manure weighs about 50 pounds.) Space the rows 14 to 18 inches apart and plant seeds at a depth of three-fourths inch. When young seedlings have sprouted undertake the first thinning. Later on, larger thinned-out plants can be cooked if the quantity is sufficient. For a continuous crop, plant new rows of spinach every week or ten days until hot weather arrives.

Care and Feeding Like all leafy crops, spinach should get a side dressing of a high-nitrogen fertilizer when plants are from six to eight inches high. Five pounds of nitrate of soda sprinkled along both sides of the row and worked into the soil with a hoe will help to produce thick abundant leaves. Weeds can be controlled either by mulching the row or by nipping them early by gently scraping the surface of the soil with a hoe.

Harvesting When picking spinach take side leaves early on. Cut the entire plant before warm weather forces it to "bolt" to seed. If your summers are on the cool side, use the resulting space for a later crop.

Insect and Diseases Sometimes blights or yellows, caused by a mosaic virus often spread by aphids, will appear on the leaves. The best cure is prevention; try to order a blight-resistant variety of spinach when sending for seeds. If aphids, beetworms, or leaf miners are evident, spray the crop with an all-purpose spray, following the recommendations in chapter 5. See also the chart in that chapter.

SQUASH, SUMMER AND WINTER

Varieties

SUMMER
Early Prolific Straightneck (yellow)
Zucchini (green)

FALL
Royal Acorn
Table Queen
Waltham Butternut

WINTER
Blue Hubbard
Buttercup
Gold Nugget

HARDINESS: *Tender*
TIME TO PLANT: *After danger of frost*
SPACING, SUMMER: *hills 4 to 6 feet*
plants 2 to 4 per hill
FALL AND WINTER: *hills 6 to 8 feet*
plants 2 to 3 per hill
DEPTH TO PLANT SEED: *½ inch*
SEEDS PER 25′: *½ ounce*
AMOUNT PER PERSON: *1 hill*
YIELD PER 25′: *37 pounds*
TIME TO HARVEST: *45 to 60 days (summer)*
85 to 115 days (fall and winter)

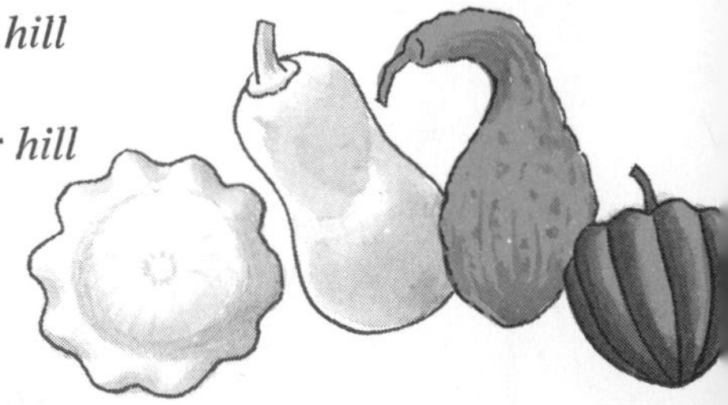

The larger vine squash may have to be passed up if garden space is limited, but there are plenty of bush varieties that can be tucked neatly in a sunny corner of any garden. Similar to pumpkin in their cultivation requirements, squash are grouped into three varieties—summer, fall, and winter. The summer varieties include crookneck and straightneck (both yellow), as well as flat-fruited white types and zucchini (green). Fall varieties are acorn and butternut. Blue Hubbard and buttercup are the common winter varieties. Just for fun, you might try one of the more unusual-looking varieties such as pattipan or Seneca Butterbar (both for eating), or Turk's Turban, an ornamental squash.

Planting Make sure the ground has warmed and the weather has settled before attempting to plant squash. In low-lying wet soils, rake or hoe the soil into hills spaced four feet apart. Sow three to five seeds per hill and thin to one or two plants per hill when seedlings have broken through. In normal soils, squash may be planted in rows or in groupings of one to three plants.

Care and Feeding If manure is readily available, give squash the benefit of generous amounts worked into the soil or placed underneath each hill (see muskmelon). A pound per hill of commercial fertilizer (5–10–10) is also helpful but doesn't do much to improve the moisture-holding qualities of your soil, so add humus or compost too. If fertilizer is applied before planting, additional amounts later won't be necessary.

Harvesting The summer varieties (crookneck, straightneck, and zucchini) should be picked when young and tender, and eaten right away. They are good only as long as a fingernail pierces the skin easily. Don't let the squash stay on the vine and become large and pulpy. Winter varieties are left to mature until rinds become thick. After harvesting it's best to leave butternut, acorn, and hubbard squash exposed to the sun or in a warm ventilated shed or garage for a few days to harden the shell before storing in the cellar where they will be protected from freezing. Ideal storage temperature is 55 to 60 degrees (see chapter 11).

Insects and Diseases Squashes can be attacked by the same insects that attack cucumber, pumpkin, and melon. For specific problems and their cures, see the chart in chapter 5.

SWISS CHARD

Varieties
Fordhook Giant
Lucullus
Rhubarb Chard

HARDINESS: *Hardy*
TIME TO PLANT: *When soil can be worked*
SPACING: *rows 30 inches*
plants 12 inches
DEPTH TO PLANT SEED: *¼ inch*
SEEDS PER 25′: *1 packet*
AMOUNT PER PERSON: *10 feet*
YIELD PER 25′: *20 plants*
TIME TO HARVEST: *60 days*

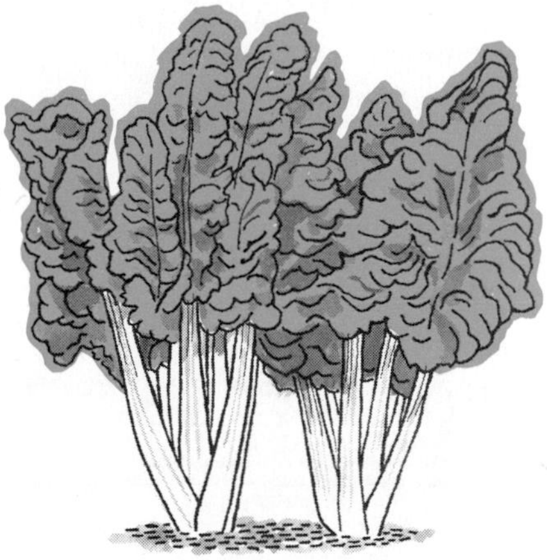

If your time is all too often gobbled up by household chores, Swiss chard is a must, because just one planting will provide continuous picking throughout the season. There is one small catch, though—Swiss chard reacts badly to an acid soil.

If a choice is possible, pick a site where the soil is mellow and rich in nutrients. Or sprinkle recommended amounts of ground dolomitic limestone over the area where chard is to be planted (see the section on pH in chapter 2).

Planting Swiss chard, which can grow to a height of from 12 to 24 inches, may be planted in rows 30 inches apart as soon as the soil can be put into decent condition for receiving seeds. But an even earlier crop can be enjoyed by starting the seeds in flats indoors three weeks before outdoor planting. Like the beet, each Swiss chard seed is actually a cluster of smaller seeds, so don't be duped into setting the seeds or plants too close together. Wide spacing will encourage stronger seedlings and facilitate thinning. A row 40 to 50 feet long will keep an average family well supplied.

Care and Feeding To get off to a good start with Swiss chard, try to work as much organic matter into the soil as possible at least two weeks before planting. Each 100 square feet of garden area could well receive three to four bushels of well-rotted manure and from four to five pounds of a balanced commercial fertilizer. Work both manure and fertilizer thoroughly into the soil and add lime as recommended above.

Harvesting Your first taste of Swiss chard will come when the row is thinned out. Pull every other plant, or however many are necessary in order to give the remaining plants at least 12 inches of space on either side. Pick the outer leaves as they become ready for use. The inner leaves will continue to grow, supplying delicious greens throughout the season. Also, you may cut off plants two to three inches above the crown and new leaves will be produced.

Insects and Diseases The same insects that are likely to invade the cabbage patch also may appear among Swiss chard plants, especially leaf-sucking and chewing insects such as cabbage green worm, plant lice, and aphids. Think twice when choosing a spray, since the leaves will eventually wind up on the family dinner table. If the worm infestation has not yet reached catastrophic proportions, try picking the pests off the plants by hand. Aphids can be effectively discouraged by scattering them with a good blast from the garden hose. If a spray becomes necessary, use a relatively nontoxic mixture such as rotenone and *do not, for any reason,* spray within three days of harvesting. See the chart in chapter 5.

TOMATOES

Varieties
NORTHERN
Campbell No. 1327
Jet Star
Spring Set
Supersonic
SOUTHERN
Floradel
Homestead
Manalucie
Rutgers
(see text for others)

HARDINESS: *Tender*
TIME TO PLANT: *After danger of frost*
SPACING: *rows 36 to 48 inches*
plants 24 to 30 inches
DEPTH TO PLANT SEED: *½ inch*
SEEDS/PLANTS PER 25′: *1 packet/10 plants*
AMOUNT PER PERSON: *5 plants*
YIELD PER 25′: *30 pounds*
TIME TO HARVEST: *75 to 85 days*

A vine-ripened tomato, still warm from the afternoon sun, is a garden treat no family should miss. Numerous variations in color, shape, and flavor make it a chameleon among vegetables. Look closely at any good seed catalog and you'll find varieties galore. All are easy to grow and delicious additions to summer menus.

Of the standard-sized tomatoes, the Rutgers variety is an old favorite. The cherry tomato—as its name implies—is small and just right for an appetizer or as a taste and color treat in salads. Pixie, Tiny Tim, and Small Fry are good cherry varieties. For making tomato paste for Italian dishes, a good variety is San Marzano. The paste tomatoes look like miniature eggplants painted red. All yellow and pink tomatoes are low in acid. Try Golden Boy or Pink-skinned Jumbo. For giant fruits that weigh over a pound apiece, choose Beefsteak or Beefeater. And then there are novelties such as the Red and Yellow Pear tomatoes, pear-shaped small fruits that are delicious as well as conversation pieces at dinner.

Try planting early, midsummer, and late varieties so that your family can enjoy sun-ripened tomatoes the whole summer.

Planting February is the time to be thinking about your tomato crops. Determine the average date of the last killing frost (see map on page 36) and plan to sow seeds in flats indoors six to eight weeks before that. When one or two sets of true leaves have appeared, transplant to individual containers. Be sure the plants spend some time in a cold frame in order to harden off; otherwise they may not survive the sudden shock of changing from indoor to outdoor life. (See chapter 3 for tips on starting plants indoors). Young plants may also be purchased from a garden center or greenhouse and may be planted directly in the garden after danger of frost has passed. If a late unexpected frost threatens, cover plants with newspaper cones, bushel baskets, or special plant protectors available at garden centers. Scrape soil against the bottom of the protectors to keep them from overturning in wind. Be sure to remove them the next morning or as soon as frost abates, to prevent plants from overheating.

Manure or other fertilizer (5–10–5) will increase the tomato yield, but both should be used sparingly since too much nitrogen can easily cause vines to run wild, producing many leaves but little fruit. Broadcast material over the entire garden rather than in rows and work thoroughly into the soil. Three to four pounds of a complete commercial fertilizer (5–10–10) or three bushels of rotted manure per 100 square feet will do. Use garden twine to keep the row straight. With a trowel or hoe, dig holes two feet apart (four feet for unstaked plants) and fill with water. When water has subsided, set plants and fill in soil. Set a three-inch collar of aluminum foil or paper around the stem and plant, making sure the collar is at least an inch below the soil (see page 44). The new soil line can be considerably higher up the plant than the old soil line. Tamp soil gently around the roots and leave a slight depression to catch rain water. Try to pick a cloudy day for transplanting to lessen the drying-out effects of a strong sun. Be sure the plants are in full sun or as much sun as possible.

Care and Feeding Tomatoes supported by stakes or specially constructed trellises not only will save space but will also produce fruit with a higher vitamin-C content as a result of increased exposure to sunlight. When plants have grown to a height of 10 to 12 inches, drive a stake into the ground four to six inches away from the plant stem. As plants mature, keep them limited to a single stem by breaking off sucker (side) shoots as they appear. Then tie the stem loosely to the stake with strips of cloth or soft twine. The plants will have to be tied two or three times during the season. Place the cloth loop just below a strong leaf branch to keep the plant from slipping down on the stake.

Supports for Tomato Plants

Tomato plants should be kept off the ground so fruit gets full exposure to sun and disease is prevented. Ways to do this are: (A) tie plants to stakes, (B) build an 8-foot long folding rack to support six plants, (C) place a flat trellis on blocks or bricks.

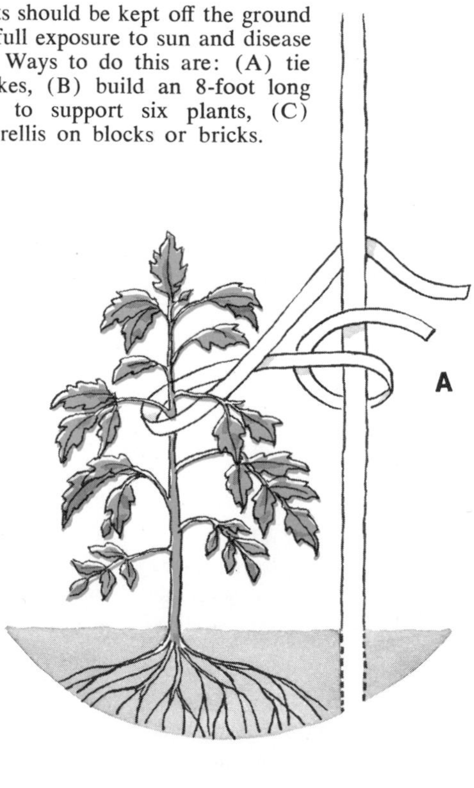

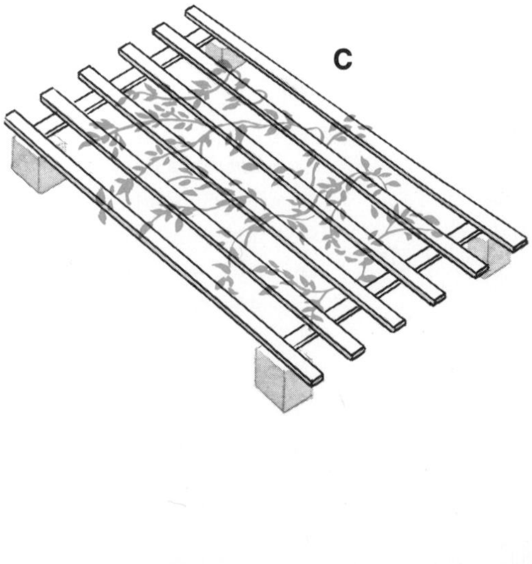

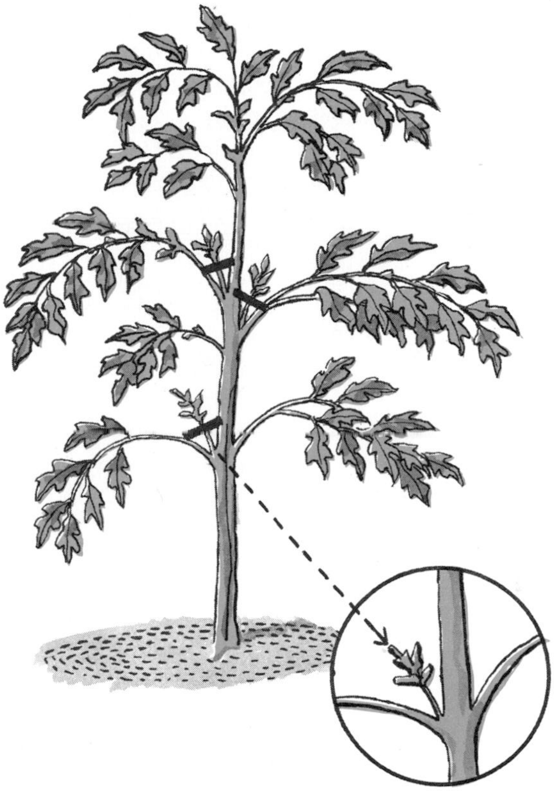

Tomato vines will produce less bush and more fruit if suckers growing where leaf stem meets the stalk are removed as they appear. Simply grasp sucker between thumb and forefinger, bend one way, then break off in the opposite direction.

In addition to plenty of nutrients and sun, tomatoes need respectable amounts of moisture. A good way to keep the moisture level of the soil high is to spread a thick layer of mulch or black sheet plastic around the plants and between the rows. Weeds will be kept to a minimum and the ground will be shielded from the direct rays of the sun. If dry weather comes, or if your area is dry, soak the garden occasionally so that the plants get around an inch or more of water per week.

Harvesting If the weather has been warm and rainfall has been abundant, tomatoes should be ready for the dinner table in 75 to 85 days, depending on variety and climate. The fruits should be firm and uniform in color. When the season comes to an end and frost seems likely, the remaining green tomatoes can be ripened either by removing them and storing them in a cool dark place (temperatures of 55 to 60 degrees) or by pulling the entire vine and suspending it upside down in a dark corner of the cellar until fruit ripens.

Insects and Diseases Tomatoes are susceptible to a host of insects, but none should reach epidemic proportions if you keep a watchful eye out for them. The first pest likely to appear is the small black or brown flea beetle that chews pinholes in the leaves. Aphids, leafhoppers, hornworms, and tomato fruitworms are all common tomato pests. As soon as plants have been set out in the garden, make a daily inspection tour. If bugs appear and are clearly causing damage, check the chart in chapter 5 for the appropriate remedy.

TURNIPS

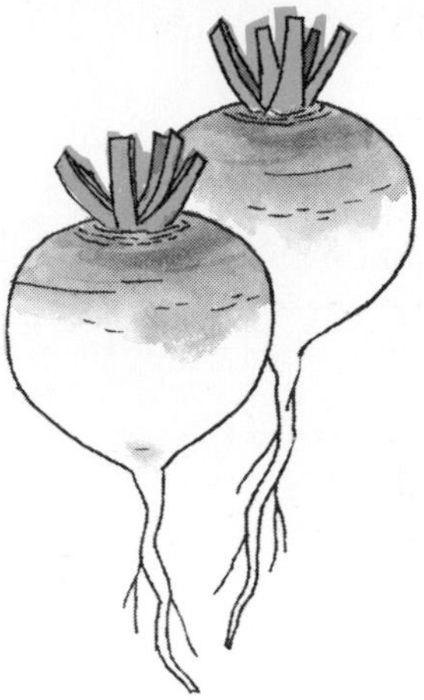

Varieties
Just Right Hybrid
Purple-top White Globe
Tokyo Cross

HARDINESS: *Hardy*
TIME TO PLANT: *Early spring or late summer*
SPACING: *rows 12 to 20 inches; plants 4 to 6 inches*
DEPTH TO PLANT SEED: *½ inch*
SEEDS PER 25′: *1 packet*
AMOUNT PER PERSON: *10 to 15 feet*
YIELD PER 25′: *35 pounds*
TIME TO HARVEST: *40 to 70 days*

A cool-season root crop that is adaptable to any section of the country, turnips can also be planted in the space of already-harvested early peas, potatoes, carrots, lettuce, or spinach.

Planting To keep roots from forking and becoming distorted, work the soil thoroughly, removing as many stones, twigs, and other obstructions as you can. If ample amounts of manure or fertilizer were applied for the earlier crop, no additional applications are necessary. If not, use a 5–10–10 formula plant food and mix in soil. Seedlings will have difficulty breaking through a soil that tends to bake or crust. Plant the seeds in a shallow furrow. Spread a combination of peat moss, sand, and soil mixed in equal amounts over the seeds to insure good germination.

Care and Feeding As soon as seedlings are identifiable, cultivate out existing weeds and then spread a mulch between the rows. When the plants are about four inches high, thin out weaker plants, leaving four to six inches in between remaining plants. Later thinnings can be cooked as greens for the dinner table. In the South this young cut-leaved foliage is used for "greens and pot likker," while the root is left in the soil to develop.

Harvesting Make sure turnips will be on the tender side by pulling them when the tops are from two to three inches in diameter. If roots are allowed to overripen they can become coarse and bitter. The roots will store well if placed in containers of sand to prevent shriveling and stored under moist conditions in temperatures ranging from 32 to 40 degrees.

Insects and Diseases Aphids are especially fond of tender turnip leaves. Spot-check the undersides of leaves occasionally and blast off any aphids with water. If they persist, use a stronger poison, carefully following the directions listed in chapter 5. See also the chart in chapter 5 for further information.

WATERMELON

Varieties
LARGE
Charleston Gray
Crimson Sweet
Dixie Queen
Fordhook
SMALL
New Hampshire Midget
Sugar Baby
Summer Festival

HARDINESS: *Tender*
TIME TO PLANT: *After danger of frost*
SPACING: *hills 8 to 10 feet*
plants 2 per hill
DEPTH TO PLANT SEED: *2 inches*
SEEDS PER 25′: *½ ounce*
AMOUNT PER PERSON: *2 to 3 hills*
YIELD PER 25′: *100 pounds*
TIME TO HARVEST: *70 to 100 days*

The chances are you won't have the necessary room in a small garden for rambling vines of watermelon (like muskmelon, planted as a vegetable but consumed as a fruit). But if space is available and your soil is on the sandy side, watermelons are well worth a try, especially for the children. In northern short-season climates, plant small quick-maturing varieties.

Planting A voracious feeder, watermelon will take all the manure or compost you can supply. When preparing hills, place two bushels of manure or compost under each mound. Then mix half a pound of commercial fertilizer with the soil. Plant at least four to five seeds and thin to the two strongest plants per hill when the seedlings have sprouted. Adaptable to a wide range of pH reactions in soil, watermelon is fussy about soil temperatures. A warm soil with good drainage is a must.

Care and Feeding Since watermelon consists mostly of water (93 percent), it is not surprising that maturing plants need abundant moisture. Water frequently in dry regions and in dry periods. Use a mulch to keep moisture in and the fruit clean.

Harvesting It might take patience and experience before you are an expert at telling when a watermelon is ready for picking. Generally the fruit is ready when the underside has turned a yellow color or when a dull hollow thud is produced by rapping it with the knuckle. If the sound is high and sharp the melon is immature.

Insects and Diseases If a wilt-resistant variety has been ordered from a reliable seedsman, the only likely threat to the crop is the striped cucumber beetle. Either pick off by hand or spray the foliage with one of the materials recommended in the chart in chapter 5.

7.

Bush Fruits and Strawberries

It's hard to imagine a backyard plot that doesn't have a few square feet of space for strawberries or a budding gardener who wouldn't walk a mile for the flavor of freshly picked raspberries in a deep dish of heavy cream. But you don't have to walk anywhere, because blackberries, blueberries, grapes, raspberries, and strawberries are all possibilities for your backyard. In fact, they are made to order and for good reasons: (1) they require minimal attention once established; (2) they offer high yields compared to their demands for space; (3) except for strawberries, they are perennial growers and will supply buckets of colorful tasty fruit season after season. But these fruits are cold sensitive. Protection from bitter winter winds is critical. And since poor planning is often the root of many garden troubles, give the berry patch extra consideration when charting your garden. Keep these factors in mind, especially if you live in the northern states:

Choose a high area over a low area, since cold tends to collect in valleys.

Purchase plants from a reliable nursery and set them out immediately to prevent roots from drying out.

Place the plantings away from the vegetable crops, if possible. Blueberries, for instance, can be worked into shrubbery borders and grapes can decorate a fence.

Provide a barrier of shrubs, trees, or fencing to block strong prevailing winter winds.

In the fall, cover crowns with a heavy mulch to guard against frost heaving.

Locate the crop on a gentle northern slope as a precaution against alternate freezing and thawing.

Keep the soil well supplied with fertilizer and moisture, especially when the fruits are forming.

There are no fruit trees (apple, peach, pear, etc.) in our plans. This is not an oversight. They have been omitted because we feel they do not fit into the average home garden. Fruit trees are better left to large gardens and estates because of space and other factors. Planting fruit trees adjacent to the vegetable garden—where they logically belong—poses certain problems. Fruit trees usually need numerous sprayings of chemical insecticides in order to keep insects and diseases at bay, and because the gardener might be intent on this (and this alone), he might forget that sprays can drift onto nearby food plants. In any case, we have offered only bush fruits and strawberries, and in chapter 6, muskmelon, pumpkin, watermelon, and rhubarb, which are not true fruits but are usually consumed as such.

BLACKBERRIES

Varieties
Bailey
Darrow
Smoothstem
Thornfree
(last two are thornless)

HARDINESS: *Winter sensitive*
TIME TO PLANT: *When soil can be worked*
SPACING: *rows 7 to 8 feet*
plants 3 to 5 feet
PLANTS PER 25′: *5*

Blackberries fall into two types, depending on the growing habits of their canes—trailing and erect. The erect blackberry develops arched, self-supporting canes, or stems, that need little or no support, while the trailing types need some kind of trellis to prevent them from becoming tangled and thoroughly unmanageable. If space is no obstacle, choose a trailing variety—many consider them better tasting.

Planting Almost any soil with good drainage will do for blackberries as long as moisture is abundant, but sandy loam with plenty of humus is the best. Be sure you locate the blackberry row within garden-hose distance of a water supply so that plenty of moisture will be available when fruits are forming. The rest is easy. Simply prepare the soil as you would for any vegetable, working in two or three bushels of rotted manure or three to four pounds of a complete fertilizer (5–10–10) for every 50 feet of row. Purchase virus-free canes from a reliable nursery and plant as soon as possible, spacing individual plants five feet apart in rows seven to eight feet apart. Set trailing types six feet apart in rows six to eight feet apart. If planting has to be delayed, keep roots moist by "heeling in"—digging a shallow trench, spreading the plants out with roots down, and covering the roots with moist soil.

Care and Feeding To keep yields heavy, spread a complete fertilizer (5–10–10) along the blackberry row every year at blossoming time. A lighter application shortly after fruiting is also a good practice. And keep weeds and grass out by weekly cultivation, being careful to keep hoeings shallow near the row where roots tend to grow close to the surface.

TRAINING—Although the erect varieties of blackberry can be grown without support, less injury and easier maintenance will result if the canes are tied with soft twine to a single wire stretched between two poles. Sink poles into the ground 15 to 20 feet apart and fasten wire to them 30 inches above the soil. Trailing varieties are handled in the same way except that two wires are used instead of one. Place one wire three feet from the ground and the other five feet. Then tie the trailing canes along the wire.

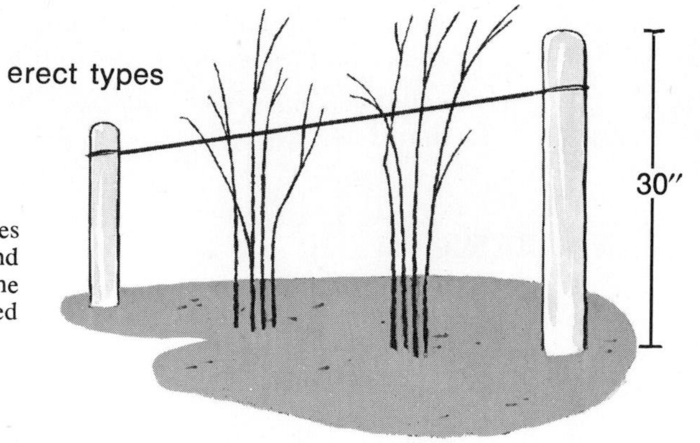

Support for both erect and trailing types of blackberries keeps canes manageable and allows easy access to fruit when picking time arrives. Wire can either be firmly twisted around poles or fastened with staples.

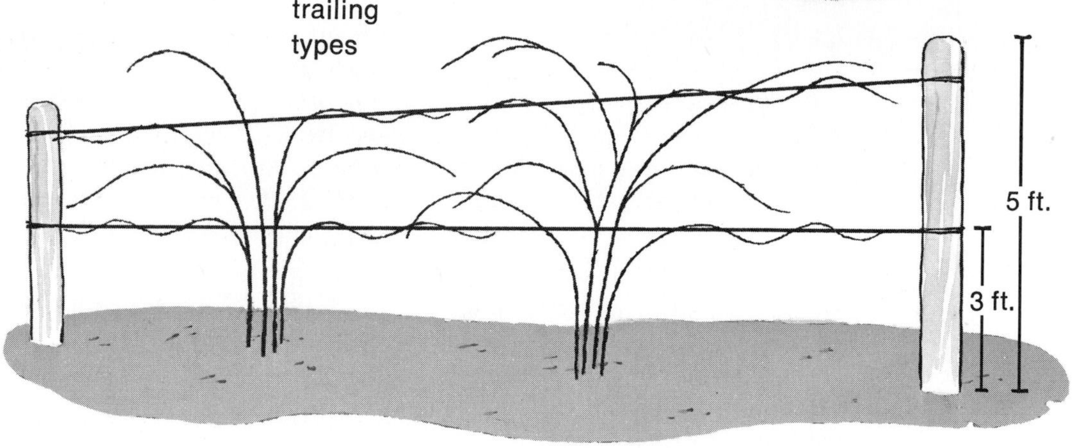

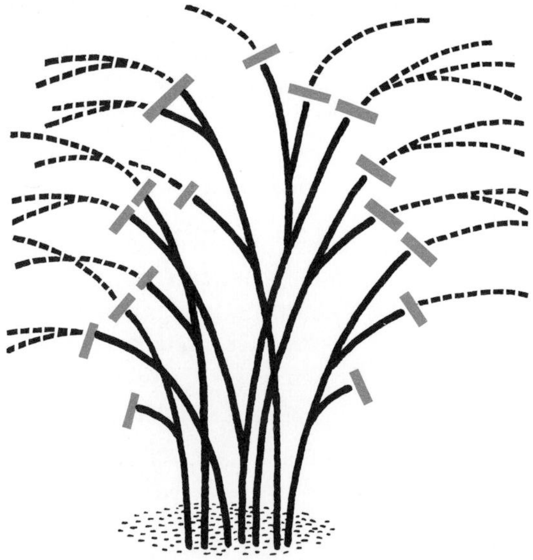

Blackberries should be pruned for greater yields. Fruiting can be improved and canes kept within manageable bounds if lateral growth is trimmed to 12 inches in length. Tops of plants should also be cut back to keep plants erect.

THINNING—Because the new canes that emerge every year from the perennial crowns beneath the soil live only two years, some thinning is required. Two stages of growth are involved. The first year the new canes send out side shoots, and during the second year the side shoots develop buds that eventually form fruits. But once fruiting has occurred the entire cane dies. Take these useless canes out of the row by periodic cutting at ground level, in either spring or fall.

PRUNING—Toward the end of the first season, the new stalks send out side branches called laterals. It is in the following season that small shoots emerge from these laterals and produce fruit. To improve the size of the fruit as well as the quantity, limit the number of fruit-producing branches by cutting back laterals until only four to six buds remain. Weaker canes should have only two buds per lateral.

Blackberries propagate themselves by sending out underground suckers into the surrounding soil. If left unchecked, your planting will become a tangled thicket. Remove suckers from between the rows as they appear. The row itself should be no more than two feet thick.

WINTER PROTECTION—In regions where temperatures are likely to plummet to 20 degrees below zero and winds often reach gale proportions, most blackberries will definitely need protection from winter injury. (Some new varieties are said to be hardier.) In the fall, before the soil freezes, bend the canes to the ground and cover them with a layer of hay, coarse manure, or soil. Then, in the spring, uncover and tie them to the wires. If winds are the main danger, not cold, provide a windbreak by building a fence or planting a row of evergreens instead of covering them up.

Harvesting A ripe blackberry will be sweet and firm. Try to pick often to prevent berries from lingering on the canes and becoming overripe. It's also a good idea to get at the picking chores early in the day before the sun is at its peak.

Insects and Diseases Although blackberries will seldom succumb to insect invasions, aphids, leafhoppers, mites, or sawflies may occasionally appear. Eliminate attractive nesting sites by pruning out infested canes and burning them. It also helps to go through the rows after harvest and cut out all old canes.

Blackberries suffer from a variety of bacterial, fungus, and root diseases. Make your garden unreceptive to disease by:

Removing and disposing of all nearby wild blackberry and raspberry plants where disease can get a start.

Keeping the garden area free of weeds and other debris.

When disease strikes, removing and destroying diseased plants.

Purchasing only healthy plants from a reliable nursery. Inspect thoroughly before you buy. Look for discolored or curled leaves, nodules or warts on roots, and rust-colored scabs on stalks.

Check the chart in chapter 5 for specific insects and diseases and their specific, effective controls.

BLUEBERRIES

Varieties
Earliblue (early)
Blueray or Rancocas (midseason)
Jersey or Coville (late)

HARDINESS: *Hardy*
TIME TO PLANT: *When soil can be worked*
SPACING: *rows 5 to 8 feet*
plants 3 to 8 feet
PLANTS PER 25′: *3 to 8*

If your soil is strongly acid and well supplied with moisture, you can count on abundant yields of blueberries two or three years after young bushes are planted. Obtain plants two or three feet high from a good nursery. Bushes with solid root balls wrapped in burlap (B & B in catalogs) will adapt much quicker and yield earlier than bushes bought with bare roots. Once established, they will require little more than a springtime pruning in the way of care. The rest is picking and the pleasure of eating fresh blueberries from the garden. Since blueberries are not self-pollinating, two or more varieties should be planted. Midseason varieties will pollinate both early and late types.

Planting The first consideration is soil acidity. Blueberries thrive on soils ranging in pH reaction from 4.0 to 5.2. But all is not lost if your soil, after testing, proves to be above this level. (See chapter 2 on how to test for pH.) By mixing acid peat moss with the planting soil in equal amounts and mulching generously with oak sawdust or acid-reacting oak leaves, ideal growing conditions can be provided. For lowering acidity with sulfur, see chapter 2.

To keep soil on the acid side, mix equal amounts of peat moss and soil when planting blueberries. A sawdust mulch will also help preserve acidity, keep weeds down, and maintain soil moisture.

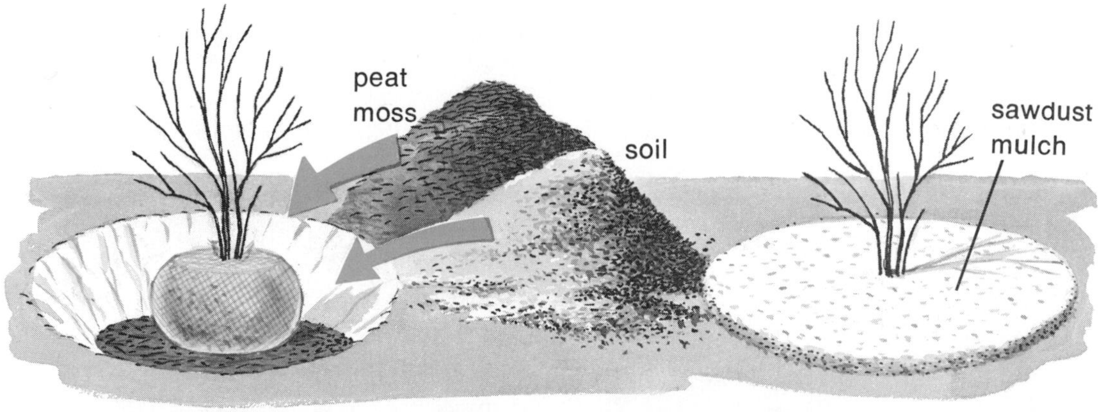

The second consideration is moisture. If a section of your property remains moist through a good portion of the year—it shouldn't flood, since too much water can drown plants—it could be an ideal location for a row of blueberry bushes. Other site requirements are good air circulation and plenty of sunlight. Try to avoid depressions where cool air gathers and frosts settle. If space is a problem, work blueberries into the shrub borders in your landscaping. Their lantern-shaped blossoms and colorful fall foliage make them a favorite of homeowners. Blueberries are very hardy. New varieties withstand 20 degrees below zero and are long-lived.

Three- or four-year-old plants should be purchased and set out as soon as possible. Dig a hole equal to twice the diameter and as deep as the root ball. Fill with water. When the water has drained, insert the plant, making sure the upper roots are no more than one to two inches below the soil surface. Although it isn't necessary to remove a burlap covering entirely, the restraining twine should be cut and the burlap loosened and rolled down. Then cover the root ball with a mixture of peat moss and soil and water thoroughly so that soil will settle, eliminating air pockets. As a final step, mulch each bush with two to four inches of sawdust, acid peat moss, or wood chips to keep weeds at a minimum and keep soil moisture from evaporating. Then remove all fruit buds, to channel all the plant's energy into growth.

Care and Feeding Relatively little care is needed, yet some attention should be given the bushes.

PRUNING—If a blueberry bush is left to its own devices, it will soon overbear, the fruit will become smaller, and new shoots will be prevented from forming. In order to keep the plant in proper balance with itself, inspect bushes carefully every spring after the fourth year. But growth habits differ depending on the variety of blueberry grown. Some kinds of bushes are open and somewhat flat-topped, while others are compact and upright in habit. The main things to remember when taking pruning shears in hand are the reasons for thinning: (1) to give the bearing shoots plenty of room in which to develop and bear fruits; (2) to allow for ample air circulation so disease won't strike; (3) to expose fruit and leaves to sunlight. Therefore cut out weak or slender stems, those that are crowding others, and large woody stems that no longer bear except near the top. It helps, too, to head back the bush slightly to force lateral growth.

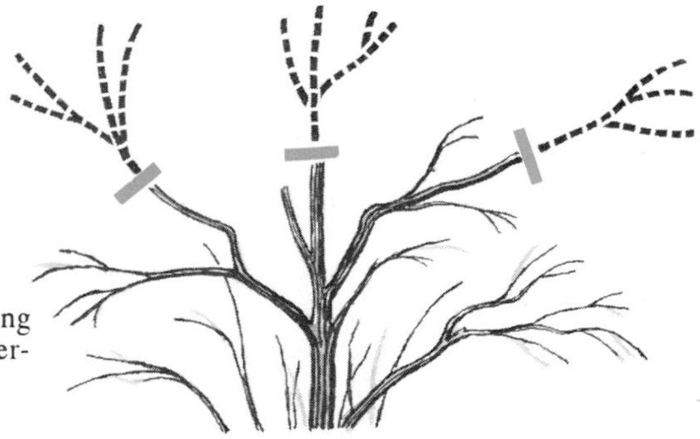

Pruning blueberry bushes in early spring before the blossoms appear prevents overbearing, which causes inferior fruit.

FEEDING—It's best not to apply fertilizer until several months after planting. In northern areas June is a good time to spread two to three ounces of ammonium sulphate (about one-half cup) in a wide circle around each plant. In subsequent years about one-fourth pound can be applied every spring as soon as you can see the buds beginning to swell. As bushes grow larger, increase the amount. A bush yielding around three to four quarts of berries can stand two applications of ammonium sulphate in doses of one-half pound each, six weeks apart, in spring. The organic equivalent is cottonseed meal. Mulch each spring to keep weeds down (as for the first planting) and to conserve moisture.

Harvesting A blue berry is not necessarily a ripe berry. It is usually two to three days after the color changes before the berries reach peak flavor and sweetness. When ready for picking they will drop readily into a bucket at a touch. If the bush is a high producer, the picking season should last two to three weeks.

Insects and Diseases Birds can be a problem. The most common defense is netting stretched over the bush. Wire netting, tobacco netting, cheesecloth, and plastic netting are materials usually available at garden centers. All will help to keep the birds from reaping the harvest right out from under you.

There are usually no insects or diseases serious enough to give the home gardener concern. Should "mummy berries" (shriveled rotting fruit) appear, consult the chart in chapter 5 for remedies.

GRAPES

Varieties
See end of this crop

HARDINESS: *Winter sensitive*
TIME TO PLANT: *When soil can be worked*
SPACING: *rows 8 to 10 feet*
plants 8 to 10 feet
PLANTS PER 25′: *2 to 3*

Grapes will need at least 170 days of frost-free weather between the last freeze in spring and the first freeze in the fall. But winter-hardy American varieties if given the proper care can produce abundantly. A good loam with plenty of humus and good drainage, located in full sun, is ideal soil for grapes. Neutral to slightly acid soil is best.

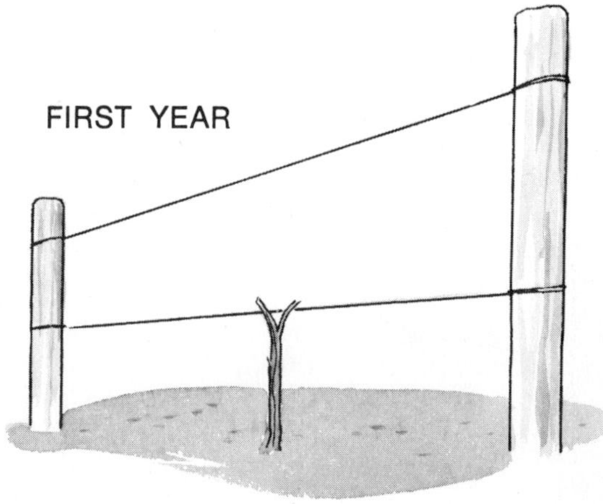

Training the grapevine to some form of trellis keeps the fruit clean, exposed to sunlight, and easy to harvest. The four-arm Kniffin system is especially favored because it requires little tying and encourages high yields.

First year: The new vine is tied with soft twine to the lower wire. Second year: When vine reaches second wire, tie again. Then select the four strongest canes as "arms" and tie along wire. Prune back all others. Third year: In addition to the four "arms," four renewal canes with two or three buds each should remain after pruning.

Planting First select a site that offers protection from strong winds and lingering frosts. Where spring weather is often unsettled, a northerly slope maintains more even temperatures than sun-drenched southern slopes. If you live near a lake or reservoir, you can take advantage of the settling influence of the water on surrounding temperatures. Protection from wind helps grapes grow and ripen well.

Purchase two-year-old vines from a reliable nursery (the Concord grape is the standard in the East against which all others are measured) and plant in the spring as soon as the soil will permit working. Cut the most vigorous canes back to two buds. All other canes should be removed entirely. Keep well watered the first season to encourage good root development.

Since ample moisture is critical, it makes sense to incorporate plenty of humus and organic material into the soil before planting. Poultry or animal manure and compost are excellent. Keep the amounts moderate, though, or overgrown vines will become susceptible to winter injury. Poultry manure is especially high in nitrogen and should be used sparingly.

Care and Feeding There are a variety of ways to train and prune grapevines, but all have the same purpose—to produce maximum yields. One of the more common methods is the four-arm Kniffin system. A central stem is trained to produce four shoots or "arms," which are tied to two wires fastened to cedar posts 18 to 20 feet apart.

After the first growing season has ended, tie the healthiest-appearing cane to the top wire and cut it off just above wire level. If the vine is not quite high enough, tie the cane to the lower wire, and by the next season the cane should be near enough to the top wire for tying. In two or three seasons, canes will be vigorous and numerous. Select four "arms," two on each side of the central cane, and tie along both wires in opposite directions. Four other strong canes should be cut back until two or three buds remain. These are termed "renewal" canes and will be the basis for new growth the following season. All other canes should be removed. When tying use a soft twine and be sure the knot is not so tight that it constricts the vine. The chief advantages of the Kniffin system are that it in-

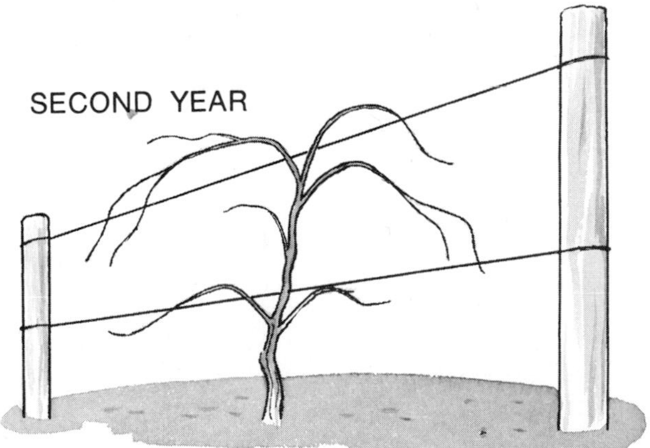

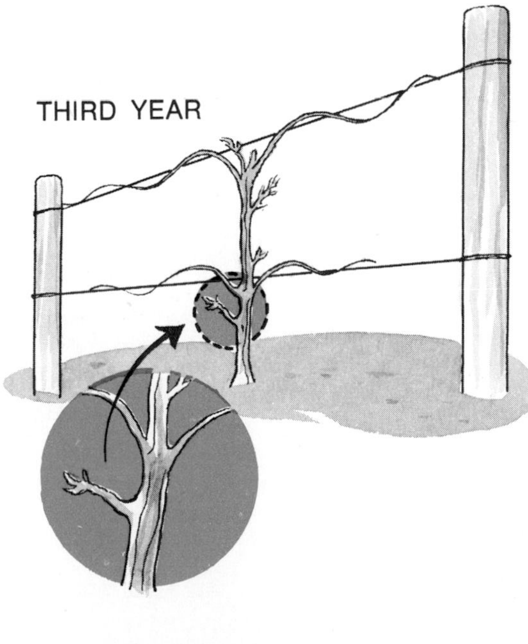

creases yields and produces clean fruit. All pruning operations should be done in late winter when the vine is dormant.

But grapes are not fussy about the kind of support they get. If space is at a premium, a vertical galvanized wire stretched between a stake in the ground and the top of a porch will do just as well. Or young vines may be held upright with a short stake or pole and canes allowed to spread at will.

Whatever system you come up with, remember that where there are severe winters grapes will survive much better if trained low to the ground where a mulch can be used to cover them. But in very humid regions the vine should be kept well off the ground to avoid fungus or mildew injury to the fruit.

Harvesting Like blueberries, you can't always tell a ripe grape by color alone. The best guide is taste. The fruit should be juicy and sweet. When picking, try to handle bunches by their stems to prevent injury. Sometimes birds will be waiting to get at the arbor. Losses can be cut down by placing plastic bags over each grape cluster; or netting material can be draped over the entire vine.

Insects and Diseases If adequate ventilation is provided and the growing season is not too humid, the grape arbor should be relatively free of diseases. Where insects or fungi are common, spraying will have to be done on a regular basis. See the chart in chapter 5 for other troubles and controls.

Reliable grape varieties and where they grow best.
NORTHEAST: **Beta, Blue Jay, Red Amber**
MIDDLE STATES: **Concord, Catawba, Delaware, Niagara**
WEST AND COASTAL STATES: **Campbell Early, Concord, Niagara, Niabell**
SOUTH: **Blue Lake, Catawba, Concord, Delaware, Niagara**
Others worth inquiring about: **Aurora, Fredonia, Caco**
And seedless varieties: **Himrod, Interlaken, Concord Seedless, Romulus**

RASPBERRIES

Varieties
See end of this crop

HARDINESS: *Winter sensitive*
TIME TO PLANT: *Spring*
SPACING: *rows 5 to 8 feet*
plants 2½ to 3 feet
PLANTS PER 25′: *8 to 10*

If your soil is deep, friable, and rich in organic matter, raspberries should definitely be given a permanent place in your garden plans. The red raspberry is the most popular, but the black and purple varieties are just as rewarding and the black are hardier in general. Some bear early, some late, while if conditions are right others may give both early and late pickings.

Planting Because raspberries need plenty of moisture, especially when the fruits are forming, the soil should contain as much organic matter as possible in order to improve its structure. The better the structure, the more moisture the soil can hold. Before planting, work in a bushel for each six plants of stable manure or compost, or add peat moss. When soil is in proper condition, obtain virus-free healthy plants from a reputable nursery and set immediately into the ground. If the roots appear at all dry, soak them in water for an hour before planting. Take care not to injure the bud at the crown in planting. Growth comes from this bud. Use a shovel or spade to force an opening in the soil; then insert roots, and close and firm the soil over them. The red raspberry can be planted two or three inches deeper than nursery depth (a close look at the stems will reveal the old soil line). Black or purple varieties should be planted at the nursery depth. Keep well watered the first season.

Care and Feeding Like blackberries, raspberries need a thorough pruning every spring. Follow this checklist when undertaking pruning chores:

1. Remove weak and spindly canes in the spring, just before buds sprout into new growth. As a rule, the thicker the cane, the more fruit it will bear. Thin stems will crowd others, robbing them of soil nutrients and preventing good air circulation, one of the best defenses against disease.

2. Keep the row width to 12 inches. As tempting as it is to let raspberry suckers grow unchecked, the row shouldn't become much thicker than 12 inches with canes six inches or so apart. Thin rows keep the fruit within reach, facilitate cultivation, and encourage the free movement of air. Suckers that spring up between the rows should be dug up and either transplanted elsewhere or destroyed.

3. Head back self-supporting canes. Most raspberries have stout enough stems to hold themselves erect, but if allowed to go unpruned they will eventually trail to the ground. With pruning shears head back black raspberry varieties to 18 to 24 inches, red and purple varieties to 30 to 36 inches. Choose a sunny day with little threat of rain in order to lessen the chance of disease invading the newly cut stem.

4. Cut back lateral growth. Toward the end of the first season the stalks send out side branches called laterals. It is the following season that small branches emerge from these laterals and produce fruit. To improve the size of the fruit as

When new raspberry plants are set out, central cane is cut back to 6 inches high. By the end of the first season, lateral shoots emerge, each bearing small buds. The second year, buds produce branches on which blossoms and fruit are formed. To improve the quality of the fruit, cut back laterals to four to six buds.

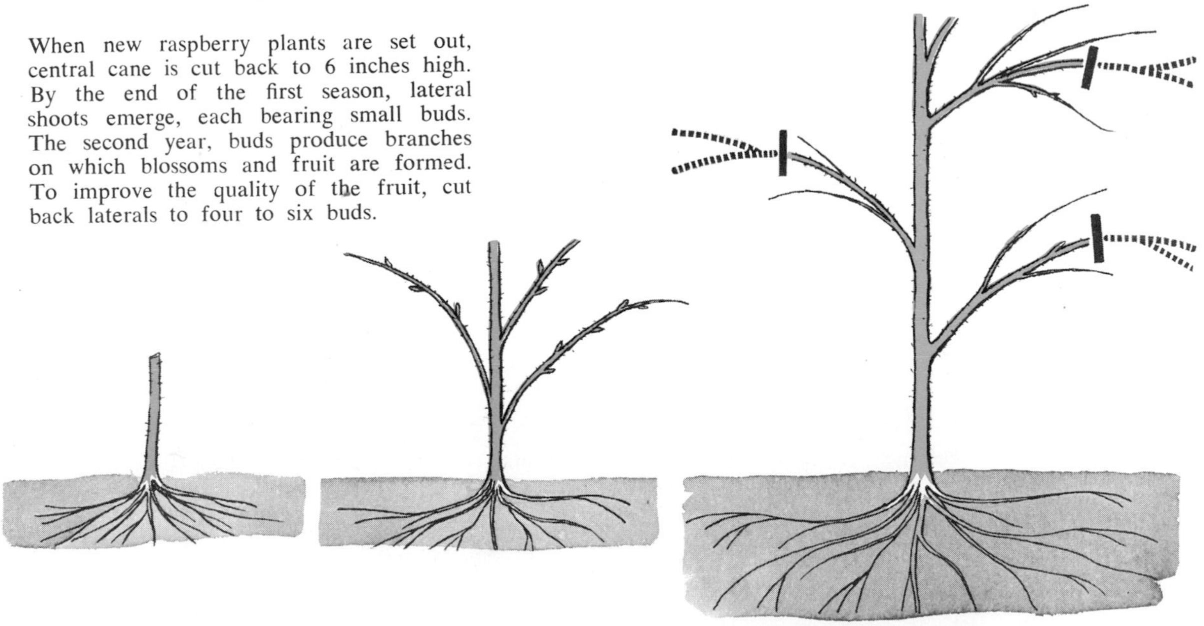

well as the quantity, it's best to limit the number of fruit-producing branches by cutting back laterals until only four to six buds remain. Weaker canes should have only two buds per lateral.

5. Remove all old spent canes after fruit has been harvested. Although crowns are perennial, new canes grow the first year, produce fruit the second year, and then die. Disease can be prevented and air circulation enhanced if these useless canes are removed and destroyed. Of course, it's easier to do things in bunches, but an excellent time to do this pruning out is just after harvest.

Raspberries need frequent cultivating to keep grass and weeds down. Be sure, though, to prevent the hoe from penetrating too deeply and injuring shallow-growing roots. Near the rows cultivation should be no deeper than two to three inches. Fall hoeings, however, can cause more harm than good, because canes may be encouraged to grow, making them more susceptible to winter injury. Mulching is an excellent way to keep the raspberry patch in good health, but if sawdust or wood chips are used, a nitrogen fertilizer will have to be applied to prevent nitrogen losses in the soil. A mulch can also come in handy if winters are severe. Simply bend the canes down to the ground and cover them with a heavy open mulch. In spring, remove the covering and allow canes to return to their normal upright position.

Like blackberries, the raspberry patch can be a much more hospitable place if a trellis or other form of support is provided. A simple arrangement consists of two 30-inch posts spaced 20 to 24 feet apart. Attach a small cross-member six inches from the top of each post to form a T. Wire fastened to ends of cross-members should be about 24 inches from the ground.

A double row of heavy wire fastened to posts will give raspberry canes all the support they need.

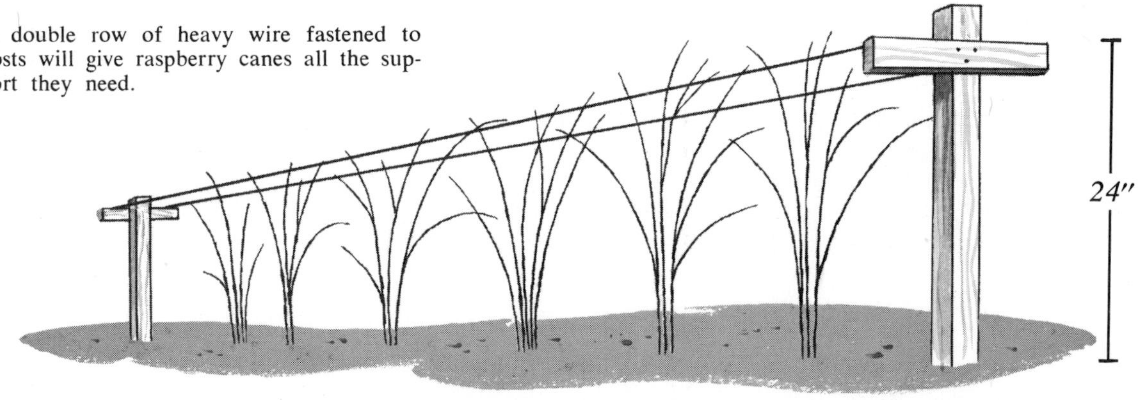

Harvesting When the berries reach maturity they ripen rapidly, especially in hot weather. Look for a deep red color on red varieties and a dull surface on other kinds. The berries should part from their stems at a touch. At the height of the season, it's best to do some picking each day, freezing those you cannot use immediately (see chapter 11).

Insects and Diseases Insects will cause little worry compared to the destructive virus diseases. See the chart in chapter 5. To keep damage at a minimum:

Get off to a good start by purchasing healthy plants from a nursery.

Rid the patch of old or diseased canes regularly.

Eliminate weeds.

Renew the bed every few years by rooting out all plants and replanting small new shoots. If virus disease shows up, dispose of all plants, and shift the new bed with new virus-free plants to another section of the garden.

Varieties
RED: **Fallred, Hilton, Indian Summer, Latham, Newburgh, New Heritage, Thornless Canby**
BLACK: **Bristol, Allen, Black Hawk, Morrison**
PURPLE: **Clyde, Amethyst, Sodus, Purple Autumn**

Check with a nursery to see whether any of these will give two crops in your area. Some do in certain regions.

STRAWBERRIES

Varieties
See end of this crop

HARDINESS: *Winter sensitive*
TIME TO PLANT: *Early spring*
SPACING: *rows 3 to 4 feet*
plants 18 to 24 inches
PLANTS PER 25′: *12 to 18*

Of the small fruits adaptable to home gardens, strawberries easily rate highest in popularity. They are easy to grow and will thrive in a wide range of soils, providing plenty of humus is present as well as good drainage. If the garden proper is already spilling over with vegetables, try strawberries as a border or in raised beds on a terrace or patio. They'll even do well growing from holes bored into a barrel filled with humus-rich soil.

Planting When casting about for a place to put the strawberry patch, look for a gentle unshaded slope facing north in order to give the plants good drainage and protection from alternating freezing and thawing in hard-winter areas. Also choose varieties known to be hardy. Temperatures will be much more consistent when the land is not exposed to direct rays of the sun. Cool moist climates are best, but strawberries will grow nearly anywhere with proper treatment. The important things are good soil, adequate moisture, and frequent cultivation.

SOIL—The best plot is one cultivated for two or more years; grubs and wireworms may inhabit recently spaded sod. About 75 percent of the strawberry's root system can be found within the upper three inches of soil. Because of this shallow growth, the bed will have to be thoroughly prepared and supplied with plenty of organic matter. Work in four to five bushels of old manure per 100 square feet (reduce by half if chicken manure is used) at least a week before planting. When planting in raised beds or in a strawberry urn (a barrel-shaped container with evenly spaced holes around the sides), mix equal amounts of compost or well-rotted manure with sand and garden soil. Once established, berries in raised beds will add color to any terrace, besides giving off a marvelous fragrance when the fruits begin to ripen.

HOW TO PLANT—Don't invite disaster by scrimping on price when purchasing plants. Search out a reliable nursery for healthy disease-free plants. Bargains, naturally, are always just around the corner, but the few extra pennies necessary for plants propagated by experts are well spent. Like other stock brought home from a nursery, strawberries should be planted right away to prevent the delicate roots from drying out. If a delay is unavoidable, "heel in" plants by digging a shallow trench and covering the roots with moist soil. If the roots are kept moist, the plants may be held up to a week in the refrigerator providing the temperature doesn't fall below 30 degrees.

Depth of planting is critical. The soil should cover the entire root but should not be allowed to smother the crown where new growth emerges. Use a wide trowel or the blade of a shovel to create a crevice in the soil. Insert plant, making sure the roots are well fanned out and the crown will not be below ground level. Remove the trowel and firmly press soil against the roots. To protect plants waiting to be set out, keep them covered with moistened burlap.

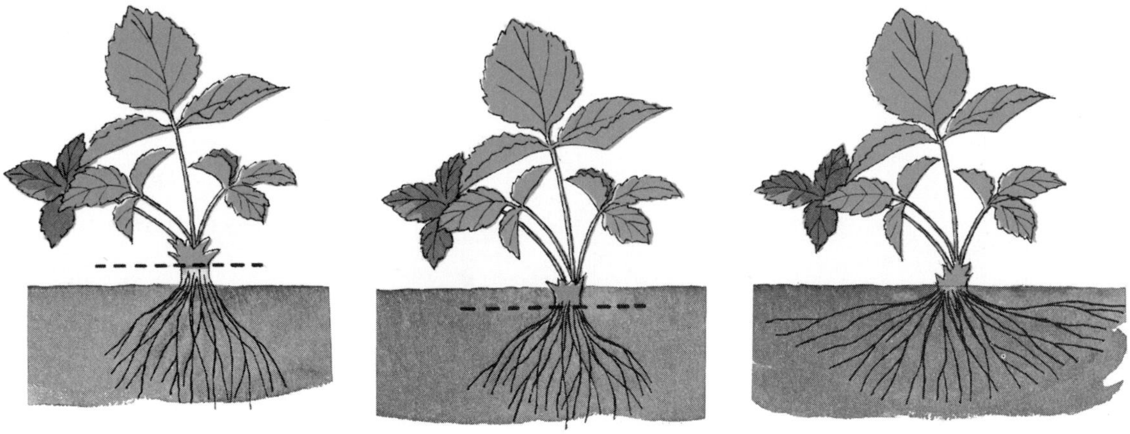

Planting depth for strawberry plants is critical. If the crown is set too high, tops of roots will dry out and the plant will do poorly even if it survives. If the crown is set too far below ground level, the plant will smother and die. Plant at right is set at proper depth, with roots well fanned out.

There are several ways to arrange the strawberry row, depending on the variety you grow as well as the kind of crop you want—either lots of small berries or a few big berries. The main difference between the growing methods is how you treat the "runners," the vinelike stems growing out from established plants at the end of which new plants develop.

Hill system: Instead of allowing plants to produce runners, the hill system limits the row to single plants spaced 12 inches apart in rows 18 inches apart. All runners are pinched off as they appear. Because of their limited growth, rows are easier to maintain. The berries are large but not as numerous. This method is preferred for everbearing varieties.

Single and double hedge row: These systems require a little more work, but are especially suitable for the home garden. "Mother plants" are set 24 inches apart in rows from 24 to 36 inches apart. The single-hedge row allows only two runners from each plant, while in the double-hedge row four runners are allowed to develop. Although extra effort is needed to be sure runners are spaced properly, the berries are large and of fine quality. Place the baby plant where you want it, staked with a clothespin till rooted. The yield is somewhat greater than the hill system but not as great as the matted-row method.

Matted row: The favorite of many commercial growers, the matted row gives runners the freedom to go where they will. Purchased plants are set out spaced 18 to 24 inches apart in rows separated by at least three feet. Let the runners root anywhere in a wide row not to exceed 24 inches in width. The advantage of this system is the increased yields, although the berries are apt to be smaller in size because of crowding.

Planting Systems for Strawberries

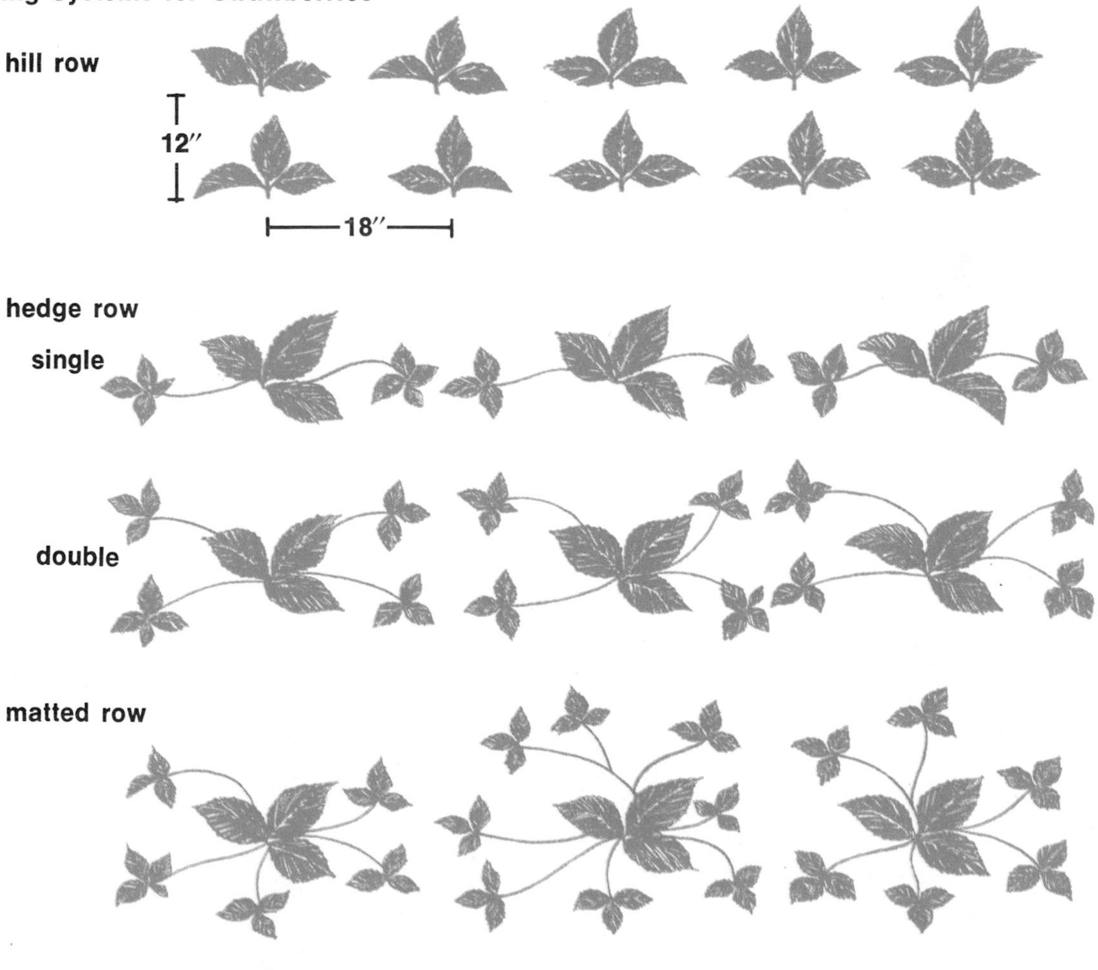

Care and Feeding Maintaining the strawberry bed involves three phases, first-season care, second-season care, and the after-harvest stage.

FIRST-SEASON CARE—Although strawberries are fast growers, they cannot be expected to establish themselves and produce an abundant crop the first year. As soon as blossoms begin to appear, pinch them off so that all plant energy goes toward building a strong root system. At the same time weeds should be kept at a minimum by going through the bed with a hoe at least every two weeks. But remember, strawberry roots stay near the surface and can be easily injured. Keep cultivation shallow. Water is another important ingredient for a successful crop. Be extra generous with water this first season, especially during dry spells and in dry climates. An inch or so a week is about right.

When fall arrives and the plants have had enough cold weather to harden them off and put a halt to growth, spread a thin mulch (about three inches) over the bed. If properly applied the mulch will cut down alternate thawing and freezing, delay growth in spring—reducing the chance of damage from a late-spring frost, maintain soil moisture, and help prevent frost penetration. Pine needles make an excellent strawberry mulch, but salt hay, marsh hay, clean straw, or even seaweed will do nicely.

SECOND-SEASON OR FRUIT YEAR—The first task in the spring is to remove the mulch as soon as you think the threat of plunging temperatures is over. Anything below 20 degrees can cause permanent damage to unprotected plants. Timing is important, since mulch left on too long will cause yellowing of the leaves. Take a look periodically to see if new growth has begun. Then pull mulch away from the row but keep it between the rows where it will preserve soil moisture and check weeds. To throttle weeds within the row and keep fruit clean, put down a thin layer of mulch between plants. Or, if mulch is spread loosely enough over the row, the plants will have no trouble growing up through the covering.

AFTER HARVEST—Generally, strawberries reach their peak yields the second season, but they will often produce a satisfactory crop the third year if the plants are kept healthy. However, if insects or diseases have plagued the bed, it's best to dig under the harvested plants and begin anew.

A good way to enjoy plenty of fresh strawberries each season is to plan three beds, one started each year. As the first patch peters out the third year, the second is producing its first crop and the third is receiving a new planting. The fourth year, the original bed is then turned under and prepared for a new planting. To get new plants, simply place a small flower pot filled with soil near a mother plant. Then take a small leaf cluster that has formed on a runner, place the cluster in the pot without cutting it free. In no time new roots will be formed. Then cut the plant free and transplant from the pot into the new location.

The exceptions to the June-yielding strawberries are the everbearing varieties, which produce fruit into late summer and fall. Although tempting because of the continuous yield, they pose a number of problems: (a) they mean more work because all runners have to be removed; (b) their early berries are inferior to the regular June-bearing varieties; (c) high temperatures of midsummer can seriously reduce the yield in many sections.

Harvesting By the unique fragrance that only ripening strawberries possess, you can tell when the bed is ready for picking. But berries are better on the firm side, so don't let them become overripe. Pick daily, and try to leave part of the stem on each fruit to conserve the juice.

Insects and Diseases Birds are likely to be the chief competitor for the strawberry crop. The only sure protection is to enclose the bed in netting as soon as berries begin to redden. Either tobacco netting or cheesecloth can be draped loosely over plants and held in place with stones or planks, or a low frame can be built to support it. For other troubles, consult the chart in chapter 5 for tips and controls.

Varieties

EARLY: **Catskill, Earlidawn, Fairfax, Red Rich, Robinson**
MIDSEASON: **Empire, Guardian, Midway, Surecrop, Red Chief**
LATE: **Jerseybelle, Sparkle**
ALL SEASON (recurrent crops): **Chief Bemidji, Dunlap, Geneva, Ogallala, Ozark Beauty, Superfection**

(When ordering, inquire about variety resistance in your area to virus and stele diseases.)

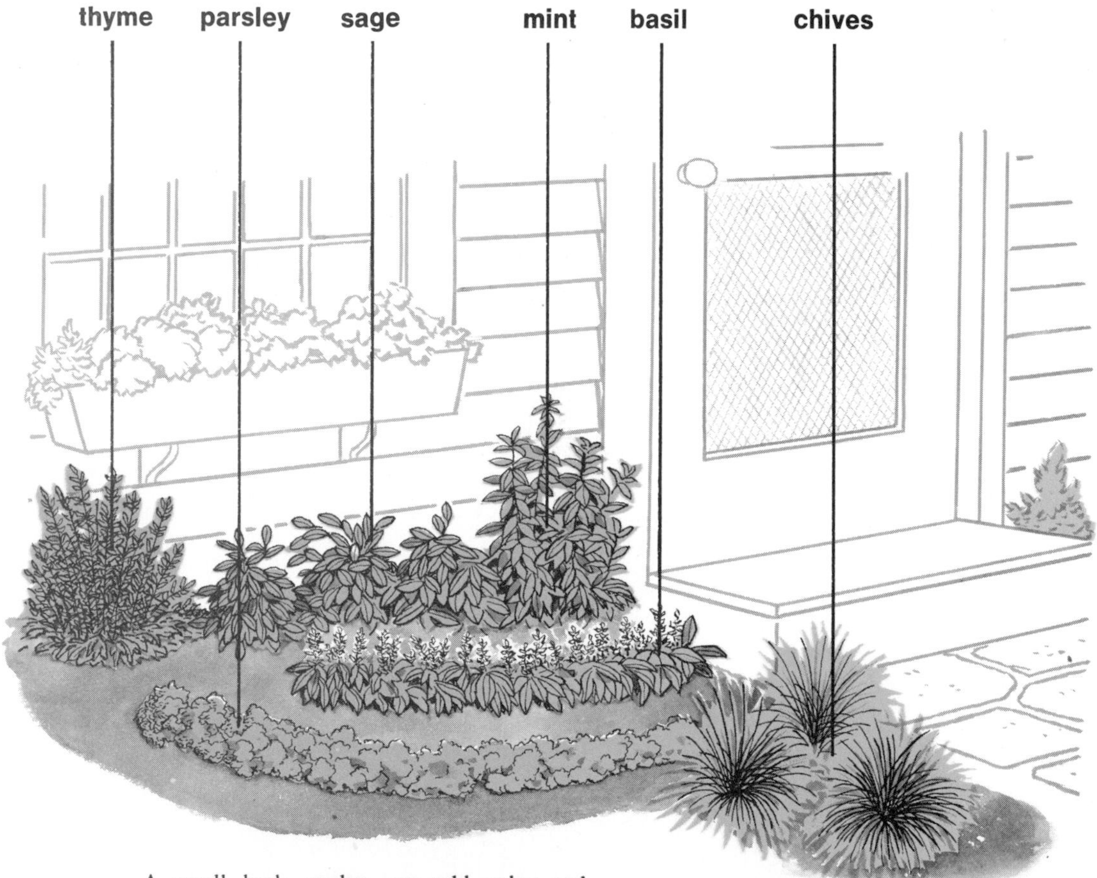

A small herb garden can add color and fragrance to your landscaping as well as delicious flavor to your meals. All herbs pictured are perennial except the rows containing parsley and basil, which must be prepared and replanted each year.

8.
Herbs and How To Use Them

Adding herbs to aroma-rich vegetables fresh from the garden might seem like gilding the lily. But herbs are the "spice" of the plant world as well as the kitchen and deserve a corner in every garden. Some need to be started each year from seed. Others will sprout every year if left undisturbed by normal garden operations. Almost all herbs require a light well-drained soil and ample sunshine. They are quite easy to grow and a delight to have and use.

There are, of course, many other herbs that the enthusiast may grow, but we have confined our list to those most useful in cooking. We feel that the average gardener will not wish to sacrifice precious garden space when other plants will bring greater and more immediate returns. For instance, plants such as caraway or cardamom are not included since their uses are limited, and they are easily found in bottles in most food stores. Chervil is not included because its uses are somewhat similar to those of parsley, but more limited. Nor do we feel that tarragon should be on this list. The tarragon offered in seed catalogs is not the gourmet's tarragon but an annual that lacks the delicate flavor one would grow it for. The *true* tarragon (often called French tarragon) is a perennial plant not widely available in this country. However, if you wish to expand your herb garden it will be well worth your while to pursue the search for tarragon plants.

To engineer early pickings of herbs, start the seeds indoors and transplant to the garden as soon as the soil has warmed (see chapter 3). Some, like mint, chives, savory, and sage, are perennials and will winter over, sprouting season after season. Others, like basil or parsley, can be transferred to small clay pots in the fall and arranged along a sunny window sill for winter use.

ANNUAL HERBS

Basil If tomatoes are grown, basil is a must. Delicious when sprinkled on freshly sliced tomatoes, bush basil has deep green or purple leaves that can almost pass for a ground cover. Plant as a border along a walk, in clumps in the garden, or put a pot or two on the terrace.

Dill If you plan to do any pickling or if you have a special weakness for salads, dill is a must. The seeds and the stems are packed right in the pickling jar along with the cucumbers. Picked fresh, the leaves add subtle flavor to salads, fish, and many other dishes. Plant dill each year, either along with vegetables or as a border plant. Remember, though, dill grows to a height of about two feet or more, so use its feathery leaves as a delicate backdrop for flowers or other shorter herbs.

Marjoram One of the most useful of herbs, marjoram produces aromatic and flavorful leaves that make a tasty addition to almost any meat dish, and are used for fish and game too. Marjoram will also do wonders for leafy vegetables, as well as for beans, peas, or squash, and as a salad-dressing additive. Start marjoram from seed either indoors early or outdoors later, but be sure the soil is warm before planting outdoors. Take advantage of its purplish pink flowers for the flower bed or as a border plant.

Parsley Parsley, the jack-of-all-trades of the herbs, can be used with virtually every vegetable as well as with a variety of meats. Technically speaking, parsley is a biennial, but it does better if planted fresh each year. To soften the hard seed cases before planting, soak seeds overnight in a cloth bag immersed in water. Sow in clumps or in rows. Be sure to harvest parsley regularly to encourage plenty of new growth.

Three types of parsley are available. There are the curled-leaf, known for its deep green color and rough mosslike texture, which make it excellent for garnishing, and the flat-leaf Italian type. Parsley can also be grown for its roots (parsnip-rooted or Hamburg variety), which are boiled and served like parsnips or included for their flavor in soups and stews. The Hamburg variety is best when roots are desired.

PERENNIAL HERBS

Chives No garden should be without a few grasslike clumps of chives. Although chives can be grown from seed, they will be enjoyed sooner if a clump is purchased, divided, and planted six to eight inches apart. The snowball-shaped lavender blossoms and wispy leaves make this herb a natural for landscaping purposes. A member of the onion family, chives have similar growing habits. Plant early in the spring as soon as the soil can be worked. If chives are a family favorite, dig up a clump or two in the fall before frost, transplant into pots, and keep on a sunny window sill for winter use.

Mint Spearmint and peppermint are the traditional varieties of mint found in most home herb gardens. Peppermint is especially popular for making drinks and flavoring jellies. Both will grow in a variety of soils, are perennial, and spread rapidly once established. Plant from seed or rooted cuttings, or obtain young plants from a nursery. Give the plants ample room to allow for eventual spreading and confine the roots as they spread invasively. As plants grow, snip off the tender tips for kitchen use. Once the plants begin to go to seed or develop woody stalks, their flavor decreases, so harvest early.

Oregano Perhaps better known as the Italian "pizza herb," oregano is the pungent cousin of marjoram and a standard addition to salads, pork, squash, and tomato dishes. Start seeds indoors in early spring and transplant to your kitchen garden when about three inches high. Because of its spreading habit, oregano is a natural for a rocky bank or along a stone walk. But make sure the soil is light in texture, well-drained, and in full sun. Take cuttings of young fresh leaves or dry whole plants before they bloom in late August (see below).

Rosemary Rosemary may be grown outdoors as a perennial in southern regions, but it is much too tender for the harsh winters of the North. In the North it makes a decorative pot plant that winters indoors. Rosemary needs a sunny but protected location—a south wall, for example. Its evergreen leaves and delicate blue blossoms make it an excellent border plant. The aromatic leaves may be used fresh or dried and are delicious on parsnips, peas, and especially squash.

Sage When it comes to poultry or game stuffings and pork or poultry dishes, sage must be within easy reach of the cook. A perennial plant requiring a rich but light soil, sage appreciates some protection from the drying effects of winter winds. Use it as a leafy backdrop for flowers or plant along a pathway where its fragrance will fill the air as you pass by. Sage can grow as high as two feet and should be cut back to encourage side growth. Purchase young plants from a nursery and set out in the garden two to three feet apart in the row. Three plants should be enough for the average family.

Once plants are established, cuttings for drying and storing can be taken two or three times per season (see below). Snip off tender young tops of plants—as much as three inches can be pruned from an eight-inch plant. Or pick leaves for immediate use.

Thyme Especially adapted to the home garden, thyme offers a variety of scents, including lemon and even caraway. Give thyme a sunny location where the soil is on the gritty side. Then plant cuttings, or plants purchased from a garden center, 20 inches apart. If an herb garden is out of the question, keep thyme in mind when looking for something to grow along a stone path or on a patio. The creeping variety, *Thymus serpyllum,* is perfect for those bare spots between flagstone steps. The stones, in turn, will preserve moisture and hold down weeds.

HARVESTING, DRYING, AND STORING HERBS

The beauty of herbs is that a world of exotic and subtle flavors and aromas can be a mere step away from the kitchen door. As soon as the plants are established, tender leaves can be snipped off selectively and sprinkled right away on salads and sliced tomatoes, or used as garnishes for meat and fish plates. Chief among the herbs grown primarily for their leaves are parsley and chives, both of which are used so regularly in gourmet cooking that they should be just outside the kitchen door. Rosemary and thyme will also supply tasty leaves, though they are prized in the garden for their blossoms too.

Keep a sharp eye out for the appearance of blossoms to determine the peak time for harvesting. When blossoms have just begun to open, make cuttings during the sunny morning hours when plant oils are at their height. Take the cuttings inside and rinse under cool water to remove dirt and insects; pick off damaged leaves. After the cuttings have dried, tie them into small bunches. Hang upside down in your kitchen, where they will give off appetizing odors and add a pleasant old-world look to the decor until used.

To dry for storage, simply spread out cuttings on a window screen, cheese-cloth, or other mesh material that will allow for the free circulation of air. Pick a dark, well-ventilated place out of the direct rays of the sun for your drying racks. A corner of the attic may be the ideal spot. Once herbs have dried and the leaves appear crisp, strip leaves from stems and store in airtight bottles to preserve the color. Keep bottles on a semidark shelf within easy reach of the family cook. To get maximum flavor, wait until dinner is ready to be served or the pot is boiling before grinding leaves for use in flavoring.

COOKING WITH HERBS

The true joy of herbs comes when bottles of dried leaves are arrayed before the cook and the pots are steaming on the stove. By artfully adding a pinch of thyme here or oregano there, you can make home-grown vegetables taste even better, whether they come directly out of the garden or from the freezer. During summer months, herbs can be used fresh from the garden when flavor is at its peak. But dried herbs, tucked away in bottles, become a goldmine of flavors that can brighten any winter day. Squash will taste the same cooked in any pot, but the cook who judiciously uses herbs can create distinctive squash dishes worthy of plush, expensive restaurants.

The key to herb cookery is moderation. After all, the main feature should be the vegetable itself. Whatever is added should *enhance* the flavor, not obliterate it. This is where your own creativity and ingenuity come in, for there are few hard-and-fast rules in herb cookery. The best cooks taste and add, taste and add as they go along. In using herbs, as in a lot of things, flavor is in the palate of the beholder.

Remember too that the delicate flavor and aroma of herbs can easily be lost if they are overcooked or allowed to dry out too much. If fresh herbs are used, chop them up or cut the leaves into small pieces to release the flavor. Dried herbs may be crumbled between the fingers as they are added. The exact moment to add herbs, of course, depends on the length of cooking time. For soups, which simmer for hours, herbs should be added about ten minutes before the pot is taken from the stove. Vegetables, on the other hand, need much less cooking time, so herbs can be sprinkled into the pot along with the vegetable. If an uncooked salad is planned, add your fresh herbs to the vinegar and let stand for at least an hour before mixing in the oil.

While amounts vary depending on individual tastes, here are some rules of thumb to get you started:

> If a recipe calls for dried herbs and you want to enjoy fresh herbs just plucked from your garden, use twice as much fresh as you would dried.

If a recipe makes no mention of herbs but you want to tinker with a dish anyway, allow:

½ teaspoon fresh herbs for each 1 pound of meat or poultry

½ teaspoon fresh herbs for every 2 cups of soup or sauce

¾ teaspoon fresh herbs for every 2 cups of vegetables

½ teaspoon fresh herbs for each 1 cup of salad dressing

Remember, some herbs, like oregano, are more pungent than milder types such as basil and can easily overpower a dish if overdone.

Herbs can only be added. If too much is used there is no way to subtract them. Go easy—add a little at a time.

See the chart that follows for suggested herb and vegetable combinations. Also see the recipes in chapter 10.

USES FOR HERBS

HERB	EGGS & CHEESE	MEAT, FISH, & POULTRY	SALADS	SOUPS	VEGETABLES
Basil	Omelets Macaroni and cheese Rarebits Scrambled eggs	Beef Chicken Duck *Fish and shellfish Lamb Veal Stuffings for fish, meat, poultry	Cucumber Egg Greens Potato Seafood Tomato French dressings	Bean Chowders Cucumber Minestrone Pea Potato Tomato Turtle Vegetable	Asparagus Brussels sprouts Carrots Eggplant Green beans Peas Potatoes Spinach Summer squash *Tomatoes Turnips
Chives	*Cottage cheese Cream cheese spread Omelets Scrambled eggs	Chicken Fish and shellfish	All salads and dressings except fruit	Garnish for all except fruit soups	All vegetables Garnish for meats, etc.
Dill	Cottage cheese Cream cheese spread Macaroni and cheese	Chicken Fish and shellfish Lamb Veal	Cole slaw *Cucumber Greens Macaroni Potato Seafood Tomato French dressings Mayonnaise *Sour cream dressings	Bean Chicken Potato Tomato	Beets Broccoli Brussels sprouts Cabbage Carrots Cauliflower Eggplant Green beans Peas Potatoes Tomatoes
Marjoram	Omelets Scrambled eggs Sharp cheese spread	Chicken Duck Fish, baked or broiled Lamb Turkey Poultry stuffings	Chicken Egg Greens Potato Seafood Tomato *Dressings	Celery Cream of chicken Onion Potato Seafood Spinach Tomato Vegetable	Carrots Eggplant Green beans Lima beans Peas Potatoes Spinach Tomatoes
Mint	Cream cheese	Fish, baked, boiled, broiled *Lamb	Cole slaw Fruit Greens Dressings for fruit	Pea	Cabbage Carrots Green beans Peas Spinach

Oregano	Deviled eggs Omelets *Pizza Sharp cheese spread	Beef Chicken Fish and shellfish Lamb Pork Veal Fish stuffings	Bean, kidney Egg Meat Tomato Vegetable Seafood aspic French dressings	Bean Chowders Minestrone Tomato Vegetable	Cabbage Eggplant Green beans Onions Potatoes Spinach Summer squash *Tomatoes
Parsley	All eggs Cheese spreads and dips Cottage cheese	Beef Chicken Fish and shellfish Lamb Veal Fish and poultry stuffings	All salads except fruit	All soups except fruit	All vegetables
Rosemary	Omelets Scrambled eggs	Chicken *Lamb Pork Fish and poultry stuffings	Chicken Greens Seafood	Chicken Fish chowders Tomato	Asparagus Green beans Peas Potatoes Summer squash Tomatoes
Sage	Cheddar Sharp cheese	Lamb and mutton Pork Poultry Veal *Poultry and fish stuffings	Chicken Turkey	Chicken Cream soups Fish chowder	Brussels sprouts Eggplant Lima beans *Onions Peas Summer squash Tomatoes
Thyme	Cottage cheese Deviled eggs Shirred eggs	Beef Chicken Fish and shellfish Pork Turkey Veal Fish and poultry stuffings	Chicken Potato Seafood Tomato French dressings	*Fish chowders Gumbo Tomato Vegetable	Beets Carrots Green beans Onions Potatoes Summer squash Tomatoes

* A particularly happy combination.

III.

After the Harvest

9.

Ways To Cook Your Crops

For the cook, the best part of home gardening comes right after the harvest. In fact, any cook worth her weight in butter knows the joy of preparing and serving fresh vegetables. Anyone can sneak out to the garden and pull up a carrot for a quick snack, but it's the family cook who has the more serious responsibility of preparing and cooking carrots so that the family receives the utmost in flavor and in nutrients at the dinner table.

COOKING FRESH VEGETABLES

It's difficult to understand why eating vegetables has become such a threat to some people. Unless, of course, the vegetables have been cooked so badly that they deserve a bad reputation. The truth is, vegetable cookery is simplicity itself. In fact, the best recipes often require the least preparation. Vegetables have marvelous flavors and lovely crisp textures, to say nothing of their beautiful colors. The job of the cook is to provide the right conditions to bring out the best in them. Once you have become familiar with different methods of cooking vegetables, don't hesitate to add your own creative touches. A sprinkling of slivered, toasted almonds or grated sharp cheese, for example, adds a bit of class to the most routine vegetable. For quicker cooking and interesting appearance, try cutting a vegetable such as asparagus or green beans on a slant instead of straight across. Or serve two together—such as tomatoes and zucchini—so their flavors and colors complement each other. There honestly is little chance of going wrong if you remember one cardinal rule: never let any added ingredient dominate or interfere with the true flavor of the vegetable.

Before you pick fruits or vegetables, check with the family gardener to see if anything has recently been sprayed. Always allow from one to seven days, depending on the type of chemical used, between spraying and eating. Wash vegetables to remove residues. Pick only as much as you plan to use for one meal, and try to harvest it as close to serving time as possible to guarantee fresh flavor. Studies have shown that the vitamin C content of most vegetables is considerably higher when they are picked late in the day, after plenty of exposure to sunshine.

Look over the produce carefully and separate any that is bruised or shows signs of decay or insect damage. Superficial blemishes are permissible—they are part of the home-gardening business—but a soft yellow cucumber, for instance, belongs on the compost pile. There are occasions when it may be more convenient to gather vegetables early in the day and cook them later. In such cases store them in a cool spot so they stay "garden fresh." Wash only if necessary and store as directed below for specific crops.

There are some specific instructions needed for cooking a few special vegetables, but these general rules may be used for all varieties:

Prepare only as much as you plan to serve. Vitamins are invariably lost when leftovers are refrigerated and reheated.

Wash thoroughly to remove sand, grit, insects, and any spray residues present. If you suspect insects may be lurking in the vegetables, soak for half an hour in a solution of 4 teaspoons salt to 1 gallon cold water. This may be particularly necessary for broccoli, Brussels sprouts, cabbage, or cauliflower.

Unless it is necessary to remove insects, do not soak vegetables in water before cooking.

If vegetables, such as carrots or potatoes, need paring, make the parings thin by using either a vegetable parer or a sharp knife. Carrots are usually scraped but may be pared thinly.

With few exceptions, cook vegetables in only a small amount of water to minimize vitamin loss.

Keep cooking time as short as possible. Vegetables should be cooked only until tender with a bit of crispness remaining. Use the timetable in this chapter as a guide, but remember the exact cooking time will depend on your altitude and on the age and variety of the vegetable and the size of the pieces. Vegetables are tender when they can be easily pierced with a fork.

Never stir vegetables unnecessarily while they are cooking.

As soon as boiled vegetables are cooked, drain them thoroughly and season to taste with butter, salt, and pepper; serve immediately. Sautéed vegetables are seasoned with salt and pepper and served right away from the skillet.

Because the flavor of garden-fresh produce shouldn't be obscured by stronger-flavored competitors, start with the following simple cooking directions for specific crops. When your hand becomes practiced and you yearn for new flavor combinations, try the suggested recipes beginning on page 166.

ASPARAGUS

Storage:

Do not wash. Store in refrigerator crisper or in plastic bag in refrigerator.

Preparation:

Break off each spear as far down as it snaps easily and discard tough ends. Remove scales with a sharp knife or stiff brush. Wash thoroughly to remove sand and grit.

Cooking:

Method 1: To skillet or large bottomed saucepan add approximately 1 inch water and ½ teaspoon salt. Bring to a boil. Add asparagus and boil, uncovered, 5 minutes. Cover tightly, reduce heat, and boil gently about 5 to 10 minutes longer or until crisp-tender. Drain thoroughly, season to taste with butter, salt, and pepper, and serve immediately.

Method 2: Stand asparagus upright in glass coffeepot, deep kettle, or bottom of double boiler. Add boiling water to cover approximately lower third of spears. Season with ½ teaspoon salt. Cover loosely with lid or upper part of double boiler, inverted. Boil gently about 10 to 20 minutes or until crisp-tender. Drain thoroughly, season to taste with butter, salt, and pepper, and serve immediately.

BEANS, GREEN or SNAP

Storage:

Store in plastic bag in refrigerator. Do not break or cut beans.

Preparation:

Wash thoroughly and break off ends. Leave whole, cut in 1-inch pieces, or French-cut lengthwise into thin strips.

Cooking:

Bring to a boil approximately 1 inch water to which ½ teaspoon salt has been added. Add beans and boil, uncovered, 5 minutes. Cover tightly, reduce heat, and boil gently until crisp-tender. Whole beans will require about 10 to 25 minutes after being covered, one-inch pieces about 5 to 15 minutes, and French-cut beans about 5 to 10 minutes. Drain thoroughly, season to taste with butter, salt, and pepper, and serve immediately.

BEANS, LIMA

Storage:

Store in pods in refrigerator crisper or in plastic bag in refrigerator.

Preparation:

Snap open pods, remove beans, and wash.

Cooking:

Bring to a boil approximately 1 inch water to which ½ teaspoon salt has been added. Add beans and boil, uncovered, 5 minutes. Cover tightly, reduce heat, and boil gently about 15 to 20 minutes longer or until tender. Drain thoroughly, season to taste with butter, salt, and pepper, and serve immediately.

BEETS

Storage:

Small young beets are best. Remove tops, leaving about 2 inches of stem (cut closely, beets will "bleed") and all of root. Store tops, following directions for GREENS, page 159. Store beets in refrigerator crisper or in plastic bag in refrigerator.

Preparation:

Wash and scrub beets thoroughly. Do not peel.

Cooking:

Place beets in saucepan. Add cold water to cover and ¾ teaspoon salt. Cover tightly, bring to a boil, reduce heat, and boil gently until tender, about 30 to 45 minutes. Drain thoroughly. Slip skins off. Slice beets or leave whole. Season to taste with butter, salt, and pepper, and serve immediately.

BROCCOLI

Storage:

Store in refrigerator crisper or in plastic bag in refrigerator.

Preparation:

Wash thoroughly and trim off outer leaves and tough bottom of stem. If stalks are larger than ½ inch in diameter, make several lengthwise slits almost to the floweret. This will allow even cooking in stalk and floweret. Or cut broccoli into 1-inch pieces. If necessary to remove insects, soak ½ hour in a solution of 4 teaspoons salt to 1 gallon cold water.

Cooking:

To skillet or large-bottomed saucepan add approximately 1 inch water and ½ teaspoon salt. Bring to a boil, add broccoli, and boil, uncovered, 5 minutes. Cover tightly, reduce heat, and boil gently about 5 to 10 minutes longer or until just crisp-tender. Drain thoroughly, season to taste with butter, salt, and pepper, and serve immediately.

BRUSSELS SPROUTS

Storage:

Store in refrigerator crisper or in plastic bag in refrigerator.

Preparation:

Wash thoroughly and remove any damaged or wilted leaves. With a sharp knife cut a small cross in stem end to allow even cooking. If necessary to remove insects, soak ½ hour in a solution of 4 teaspoons salt to 1 gallon cold water.

Cooking:

Bring to a boil 1 inch water to which ½ teaspoon salt has been added. Add Brussels sprouts and boil, uncovered, 5 minutes. Cover tightly, reduce heat, and boil gently about 10 to 15 minutes longer or until tender. Drain thoroughly, season to taste with butter, salt, and pepper, and serve immediately.

CABBAGE, GREEN

Storage:

Store in refrigerator crisper or in plastic bag in refrigerator.

Preparation:

Wash thoroughly. Remove any imperfect or wilted leaves before preparing to shred or cut in wedges.

Shredded: Cut cabbage in quarters and remove core. Place cabbage, cut side down, on firm surface and slice closely to make thin shreds.

Wedges: Cut cabbage in 6 to 8 wedges. Remove most of core, leaving just enough to keep wedges intact.

Cooking:

Shredded: Bring to a boil approximately ½ inch water to which ½ teaspoon salt has been added. Add shredded cabbage and boil, covered, until limp, about 1 minute. Remove cover and stir gently. Cover and boil gently until crisp-tender, about 3 to 10 minutes. Drain thoroughly, season to taste with butter, salt, and pepper, and serve immediately.

Wedges: Bring to a boil approximately 1 inch water to which ½ teaspoon salt has been added. Add cabbage and boil, uncovered, 5 minutes. Cover tightly, reduce heat, and boil gently about 5 to 10 minutes longer or until crisp-tender. Drain thoroughly, season to taste with butter, salt, and pepper, and serve immediately.

CABBAGE, RED

Follow directions for green cabbage. Add 1 tablespoon lemon juice to water.

CARROTS

Storage:

Remove tops to prevent wilting. Store carrots in refrigerator crisper or in plastic bag in refrigerator.

Preparation:

Wash and scrape or pare thinly. If carrots are young, they do not need paring. Leave whole or cut as desired in slices, cubes, or strips.

Cooking:

Bring to a boil approximately 1 inch water to which ½ teaspoon salt has been added. Add carrots and cover tightly. Return to a boil, reduce heat, and boil gently until crisp-tender. Cook whole carrots about 15 to 25 minutes, slices or cubes about 10 to 20 minutes, and strips about 10 to 15 minutes. Drain thoroughly, season to taste with butter, salt, and pepper, and serve immediately.

CAULIFLOWER

Storage:

Store in refrigerator crisper or in plastic bag in refrigerator.

Preparation:

Cut off woody base and remove any tough leaves. Small tender leaves may be left on the head or removed and served in salad. Wash thoroughly. If necessary to remove insects, soak ½ hour in a solution of 4 teaspoons salt to 1 gallon cold water. Leave head whole or separate blossom into clusters.

Cooking:

Bring to a boil approximately 1 inch water to which ½ teaspoon salt has been added. Add cauliflower and cover tightly. Return to a boil, reduce heat, and boil gently until crisp-tender. Cook whole head about 15 to 25 minutes, clusters about 8 to 15 minutes. Drain thoroughly, season to taste with butter, salt, and pepper, and serve immediately.

CORN

Storage:

Store, unhusked, in refrigerator.

Preparation:

Just before cooking, remove husks and silk. Cut off any sections of cob that are discolored or have been damaged in any way.

Cooking:

In a large pot, bring to a boil enough water to cover corn. Do not add salt or corn will be toughened. Add corn and cover tightly. Return to a boil, reduce heat, and boil gently 5 to 15 minutes, depending on size and age of ears. Remove from pot with tongs and serve immediately with butter, salt, and pepper.

EGGPLANT

Storage:

Wash and dry. Do not pare. Store in refrigerator crisper or in plastic bag in refrigerator.

Preparation:

Pare, if desired, and cut into ½-inch slices.

Cooking:

Dip slices first into milk, then into seasoned flour or crumbs. Sauté in 2 to 3 tablespoons butter or butter and olive oil; turn, sauté other side till a crisp golden crust is achieved. Season with salt and pepper and serve immediately.

GREENS (Swiss chard, beet tops, turnip tops)

Storage:

Remove and discard tough stems and any imperfect leaves. Wash leaves several times in a large amount of lukewarm water. After each washing, lift greens gently out of water to let sand and grit settle to bottom. Drain thoroughly and store in refrigerator crisper or in plastic bag in refrigerator.

Preparation:

Rinse in cool water.

Cooking:

If greens are young, they may be cooked merely in the water that clings to their leaves and no additional water is necessary. Otherwise, bring to a boil approximately ½-inch water to which ½ teaspoon salt has been added. Add greens, cover tightly, and cook until wilted (about 1 minute). Remove cover for a moment to let steam escape. Re-cover and boil gently until tender—for Swiss chard about 10 to 20 minutes, beet tops about 5 to 15 minutes, and turnip greens about 10 to 30 minutes. Drain thoroughly, season to taste with butter, salt, and pepper, and serve immediately.

LEEKS

Storage:

Store in refrigerator crisper or in plastic bag in refrigerator.

Preparation:

Cut off the ends of the roots and all but 1 inch of the green stalks. Peel the filmy skin off the white part, cut stalks in half lengthwise and wash thoroughly under running water, making sure no grit or sand is left.

Cooking:

Bring to a boil enough water to cover leeks to which 1 teaspoon of salt has been added. Add leeks and cook uncovered for 10 to 15 minutes, depending on the thickness of the leeks. Test for tenderness by piercing green stalks with a fork. Drain thoroughly. Season with butter, salt, and pepper, and serve immediately.

ONIONS

Storage:

Store onions in a loosely woven or open mesh bag in a cool, dry, well-ventilated place.

Preparation:

Cut thin slice from stem and root ends. Hold onion under cold running water to prevent tears and peel off outside skin.

Cooking:

Bring to a boil approximately 1 inch cold water to which ½ teaspoon salt has been added. Add onions and cover. Return to a boil, reduce heat, and boil gently until tender. Lift lid 3 or 4 times while cooking to let steam escape. Boil small whole onions about 15 to 25 minutes and larger onions about 20 to 40 minutes. Drain thoroughly. Season with butter, salt, and pepper. Serve immediately.

PARSNIPS

Storage:

Remove tops to prevent wilting. Store parsnips in refrigerator crisper or in plastic bag in refrigerator.

Preparation:

Wash and scrape or pare thinly. Leave whole or cut in slices, cubes, or strips. If parsnips have large, woody cores, remove cores with a sharp knife.

Cooking:

Bring to a boil approximately 1 inch water to which ½ teaspoon salt has been added. Add parsnips and cover tightly. Bring to a boil, reduce heat, and boil gently until crisp-tender. Cook whole parsnips about 20 to 40 minutes, slices or cubes about 15 minutes, and strips about 8 to 15 minutes. Drain thoroughly. Season with butter, salt, and pepper, and serve immediately.

PEAS

Storage:

Store in pods in refrigerator crisper or in plastic bag in refrigerator.

Preparation:

Just before cooking, snap open pods, remove peas, and wash.

Cooking:

Bring to a boil approximately 1 inch water to which ½ teaspoon salt has been added. Add peas and boil, uncovered, 5 minutes. If peas are very young, they may be cooked at this point. If they are not tender yet, cover pan tightly, reduce heat, and boil gently about 3 to 5 minutes longer or until tender. Drain thoroughly, season to taste with butter, salt, and pepper, and serve immediately.

GREEN PEPPER

Storage:

Wash and dry. Store in refrigerator crisper or in plastic bag in refrigerator.

Preparation:

Cut peppers open and remove seeds, core, and white membrane. Cut into strips ½ inch wide.

Cooking:

Sauté in butter or butter and olive oil, about 1 tablespoon per pepper. Cook until crisp-tender, about 10 to 15 minutes. Stir to prevent sticking and uneven cooking. Season to taste with salt and pepper, and serve immediately.

POTATOES

Storage:

Store in a cool, dry, dark, well-ventilated place.

Preparation:

Scrub with a soft brush. If potatoes are to be cooked whole, do not pare; new potatoes, especially, should be boiled in their jackets. If they are to be sliced or cut up, pare thinly and cut in quarters or cubes. Large potatoes may be halved or quartered to insure even cooking. Potatoes may be held for a short time before cooking by covering with cold water to prevent darkening. Avoid long soaking.

Cooking:

Bring to a boil approximately 1 inch water to which ½ teaspoon salt has been added. Add potatoes and cover tightly. Return to a boil, reduce heat, and boil gently until tender. Cook whole potatoes about 25 to 40 minutes (new potatoes about 15 to 20 minutes), quartered potatoes about 20 to 25 minutes, and cubed potatoes about 10 to 15 minutes. Drain thoroughly, season to taste with butter, salt, and pepper, and serve immediately. Garnish new potatoes with fresh chopped parsley, if you wish.

SPINACH

Storage:

Remove and discard tough stems and any imperfect leaves. Wash leaves several times in large amount of warm water. After each washing, lift leaves gently out of water to let sand and grit settle to bottom. Drain thoroughly and store in refrigerator crisper or in plastic bag in refrigerator.

Preparation:

Rinse in cool water.

Cooking:

Place rinsed spinach in large pot with only the water that clings to the leaves. Do not add any more water. Sprinkle with ½ teaspoon salt. Cover tightly and cook 3 to 10 minutes or until wilted and tender. Drain thoroughly, season to taste with butter, salt, and pepper, and serve immediately.

SQUASH, SUMMER

(see also Zucchini, next page)

Storage:

Wash and dry. Store in refrigerator crisper or in plastic bag in refrigerator.

Preparation:

Remove any damaged spots. Cut off thin slice at stem and blossom ends. Do not pare. Cut into slices ¼ to ½ inch thick.

Cooking:

Bring to a boil approximately ½ inch water to which ½ teaspoon salt has been added. Add squash and cover tightly. Bring to a boil, reduce heat, and boil gently until tender but firm, about 8 to 15 minutes. Drain thoroughly, season to taste with butter, salt, and pepper, and serve immediately.

SQUASH, WINTER

Storage:

Store in a dry cool place at about 60 degrees.

Preparation:

Remove stem. Cut squash into slices about 2 inches thick and then cut slices into quarters. Pare thinly and remove seeds and stringy center portion.

Cooking:

Bring to a boil approximately 1 inch water to which ½ teaspoon salt has been added. Add squash and cover tightly. Bring to a boil, reduce heat, and boil gently until tender, about 15 to 25 minutes. Drain thoroughly, season to taste with butter, salt, and pepper, and serve immediately.

SWISS CHARD, See GREENS, page 159

TURNIPS

Storage:

Remove tops. Save and store them following directions for GREENS, page 159. Store turnips in refrigerator crisper or in plastic bag in refrigerator.

Preparation:

Wash turnips and pare thinly. If turnips are small they may be left whole; if not, cut in ¼-inch slices or ½-inch cubes.

Cooking:

Bring to a boil approximately 1 inch water to which ½ teaspoon salt has been added. Add turnips and cover tightly. Bring to a boil, reduce heat, and boil gently until tender. Cook whole turnips about 20 to 30 minutes, slices about 9 to 12 minutes, and cubes about 10 to 20 minutes. Drain thoroughly, season to taste with butter, salt, and pepper, and serve immediately.

ZUCCHINI

Storage:

Wash and dry. Store in plastic bag in refrigerator or in refrigerator crisper.

Preparation:

Cut off thin slice at stem and blossom ends. Do not pare. Cut into slices ¼ to ½ inch thick.

Cooking:

Boiling: Bring to a boil approximately ½ inch water to which ½ teaspoon salt has been added. Add zucchini and cover tightly. Bring to a boil, reduce heat, and boil gently until tender but firm, about 8 to 15 minutes. Drain thoroughly, season to taste with butter, salt, and pepper, and serve immediately.

Sautéeing: Heat enough butter or butter and olive oil to cover bottom of skillet, about 2 tablespoons per each pound of zucchini. Add slices and sauté gently, turning slices until slightly golden and tender but not mushy. Zucchini cooks very quickly, usually in less than 8 minutes. Season with salt and pepper, and serve immediately.

PRESSURE COOKING

By no means is vegetable cookery limited to boiling. Vegetables can be cooked lots of ways; it all depends on your equipment and imagination. Pressure cooking is awfully good for vegetables because it keeps cooking times short and thereby preserves vitamins. Follow the manufacturer's directions, but keep in mind that tender young vegetables may require even less time than is suggested.

BAKING

Baking is a handy way to cook some vegetables and is particularly convenient and economical if other foods can be baked at the same time.

1. Prepare the vegetable according to the directions on pages 155–163. Arrange in an ovenproof casserole.
2. Add just enough water to cover bottom of casserole. Season with salt and pepper and dot with butter.
3. Cover tightly and bake in a preheated moderate oven (375 degrees) until tender. If the oven has been set higher or lower, adjust the baking time accordingly and cook until tender.

BAKING TIMETABLE

Carrots, whole	35 to 45 minutes
sliced	30 to 40 minutes
Onions, large, whole	50 to 60 minutes
Parsnips	30 to 45 minutes
Potatoes, medium, whole	60 to 70 minutes
Squash, summer, ½ inch pieces	40 to 50 minutes
Squash, winter, 2 inch pieces	40 to 60 minutes
Tomatoes	15 to 30 minutes
Turnips, cubed	40 to 50 minutes

PAN OR SKILLET COOKING

Here's a speedy way to cook some vegetables. It's fast because the vegetables are shredded or cut in small pieces. Then they are cooked in a tightly covered skillet with butter and a small amount of water.

1. Prepare the vegetable according to the directions on pages 155–163.
2. Melt about 2 tablespoons butter in a large heavy skillet.
3. Add the prepared vegetable and season with salt and pepper. Add about 1/4 cup water.
4. Cover tightly and cook over low heat, stirring occasionally, until tender. Add more water if vegetable begins to stick.

SKILLET COOKING TIMES

Asparagus, cut diagonally in pieces 1½ inches long	5 to 6 minutes
Beans, Green or Snap, cut in pieces 1 inch long	20 to 25 minutes
Cabbage, finely shredded	6 to 8 minutes
Carrots, thinly sliced	10 minutes
Corn, cut off the cob	15 to 18 minutes
Spinach, finely shredded	6 to 8 minutes
Summer squash, thinly sliced	12 to 15 minutes

10.
A Garden of Recipes

Once your fresh vegetables have held the mealtime limelight and have been served without garnishes or herbs, you'll soon grow restless for more exotic ways to prepare home-grown food. In no time you'll discover many variations on a single cooking theme—vegetables that complement other vegetables, sauces that turn ordinary dishes into Saturday-night specials, or a fruit dessert that will make the family gardener wish he had planted 200 rows of strawberries instead of two. And once produce begins to pour into the kitchen from your backyard cornucopia, you won't hesitate to cast about for new ways to prepare vegetables and fruits.

The important thing is to capitalize on the flavor of the particular vegetable or fruit you are using. Nothing will discourage the gardener quicker than a vegetable dish boiled to an unrecognizable mush. And if the vegetable as well as its flavor goes up in steam, you might as well stick to bland supermarket produce.

In judging and preparing the following recipes we have asked ourselves, "Are all the wonderful color, flavor, and texture qualities of garden-fresh vegetables and fruits still present by the time they get to the dinner table?" Sometimes sauces, herbs, or pan-frying will lead to remarkable flavor combinations, but the subtle squash may be so obscured you could easily have substituted sliced sponges. Try to preserve the shapes, textures, and colors that make home-produced vegetables and fruits so rewarding. Keep in mind that cooking times for all vegetables should be kept to a minimum so they do not lose color, shape, or nutritive value. And sauces should be used cautiously so they do not hide or overpower the flavor of broccoli, cauliflower, or beets. Make the recipe serve the vegetable instead of vice versa and every dish will be a masterpiece.

Asparagus, Ham, and Cheese Sandwiches

24 spears asparagus
½ teaspoon salt
4 slices bread
4 slices boiled ham
4 slices Swiss cheese

1. Break off each asparagus spear as far down as it will snap easily. Remove scales with a sharp knife. Wash thoroughly to remove sand and grit. Keep spears whole; do not cut up.

2. In a skillet or large-bottomed saucepan bring to a boil approximately 1 inch water and the salt. Add asparagus and boil, uncovered, 5 minutes. Cover tightly, reduce heat, and boil gently about 5 to 10 minutes longer or until crisp-tender. Drain thoroughly.

3. While asparagus is cooking, toast bread and arrange on a baking sheet or individual heat-proof serving plates. Place 6 spears of cooked asparagus on each slice of toast. Top each with a slice of ham and a slice of cheese. Place under a preheated broiler and broil until cheese melts. Serve immediately.

Yield: 4 servings

Asparagus and Scallions, Chinese Style

36 spears asparagus
2 tablespoons salad oil
⅔ cup thinly sliced scallions, including green tops
¼ cup water
1 teaspoon cornstarch
½ teaspoon sugar
⅛ teaspoon pepper
1 tablespoon soy sauce
2 tablespoons cold water

1. Break off each asparagus spear as far down as it will snap easily. Remove scales with a sharp knife. Wash thoroughly to remove sand and grit. Cut diagonally in pieces about 1½ inches long, cutting so the tips are left whole.

2. In a large skillet heat the salad oil. Add asparagus and sliced scallions, and cook over moderate heat, stirring gently, 2 minutes.

3. Add the ¼ cup water. Cover and cook about 4 minutes longer or until asparagus is crisp-tender.

4. In a small custard cup combine cornstarch, sugar, pepper, soy sauce, and cold water. Mix until cornstarch is dissolved. Add to asparagus and cook, stirring gently, for a few moments or until asparagus is glazed. Serve immediately.

Yield: 6 servings

Asparagus and Tuna au Gratin

36 spears asparagus
½ teaspoon salt
¼ cup butter
3 tablespoons flour
2 cups hot milk
½ teaspoon salt
¼ teaspoon pepper
1 tablespoon lemon juice
1 cup grated sharp Cheddar cheese
2 cans (7 oz. each) tuna
1½ cups soft bread crumbs
¼ cup butter

1. Break off each asparagus spear as far down as it will snap easily. Remove scales with a sharp knife. Wash thoroughly to remove sand and grit. Do not cut up spears.

2. In a large skillet or large-bottomed saucepan bring to a boil approximately 1 inch water and ½ teaspoon salt. Add asparagus and boil, uncovered, 5 minutes. Cover tightly, reduce heat, and boil gently about 5 to 10 minutes longer or until crisp-tender. Drain thoroughly and arrange in shallow baking dish.

3. While asparagus is cooking, make sauce. In a medium saucepan, melt ¼ cup butter. Using a wire whisk, blend in flour. Add milk and cook over moderate heat, stirring constantly, until thickened and smooth. Boil and stir 2 minutes. Blend in remaining ½ teaspoon salt, pepper, lemon juice, and cheese.

4. Drain tuna and break into bite-size pieces. Arrange over asparagus. Cover with sauce. Top with crumbs and dot with remaining ¼ cup butter.

5. Bake in a preheated moderate oven (350 degrees) 25 minutes, or until crumbs are lightly browned and mixture is hot and bubbling.

Yield: 6 servings

Green Beans Amandine

½ teaspoon salt
5 cups green beans (about 1 pound), French-cut lengthwise in thin strips
3 tablespoons butter
⅓ cup slivered blanched almonds
¼ teaspoon salt
⅛ teaspoon pepper

1. In a saucepan, bring to a boil approximately 1 inch water to which the ½ teaspoon salt has been added. Add beans and boil, uncovered, 5 minutes. Cover tightly, reduce heat, and boil gently 5 to 10 minutes longer or until crisp-tender. Drain thoroughly and return beans to pan.

2. While beans are cooking, melt butter in a small skillet. Add almonds and cook over low heat, stirring frequently until lightly browned. Season with remaining ¼ teaspoon salt and the pepper. Pour over hot cooked beans and toss lightly to blend. Serve immediately.

Yield: 4 to 6 servings

Green Beans with Buttered Crumbs

½ teaspoon salt
5 cups green beans (about 1 pound), cut diagonally in 1½-inch pieces
¼ cup butter
1⅓ cups soft bread crumbs
¼ cup grated Parmesan cheese
3 tablespoons chopped parsley

1. In a saucepan, bring to a boil approximately 1 inch water to which the salt has been added. Add beans and boil, uncovered, 5 minutes. Cover tightly, reduce heat, and boil gently 5 to 10 minutes longer or until crisp-tender. Drain thoroughly.

2. While beans are cooking, melt butter in a small skillet. Add bread crumbs and cheese; mix lightly to blend. Cook over low heat, mixing gently, until lightly browned.

3. Sprinkle beans with crumb mixture and garnish with parsley. Serve immediately.

Yield: 4 to 6 servings

Green Bean Salad

½ teaspoon salt
5 cups green beans (about 1 pound), French-cut lengthwise in thin strips
⅓ cup thinly sliced scallions, including green tops
¼ cup chopped parsley

DRESSING:
3 tablespoons olive oil
1 tablespoon lemon juice
½ teaspoon sugar
¼ teaspoon salt
⅛ teaspoon pepper

1. In a saucepan, bring to a boil approximately 1 inch water to which the ½ teaspoon salt has been added. Add beans and boil, uncovered, 5 minutes. Cover tightly, reduce heat, and boil gently 5 to 10 minutes longer or until crisp-tender. Drain thoroughly.

2. Add scallions and parsley to hot beans and toss gently to blend. Cool to room temperature.

3. In a cup mix together olive oil, lemon juice, sugar, the ¼ teaspoon salt, and pepper. Add to bean mixture and toss gently. Refrigerate and serve chilled.

Yield: 4 to 6 servings

Lima Beans with Sour Cream

½ teaspoon salt
2 cups shelled lima beans
1 tablespoon butter
¼ teaspoon salt
¼ teaspoon Italian seasoning
⅛ teaspoon pepper
⅓ cup sour cream

1. In a saucepan, bring to a boil approximately 1 inch water to which the ½ teaspoon salt has been added. Add beans and boil, uncovered, 5 minutes. Cover tightly, reduce heat, and boil gently about 15 to 20 minutes longer or until tender. Drain thoroughly.

2. While beans are cooking, melt butter in another saucepan. Blend in remaining ¼ teaspoon salt, Italian seasoning, and pepper. Add cooked beans and sour cream to butter and seasonings. Cook over low heat, mixing gently, until hot. Do not boil. Serve immediately.

Yield: 4 servings

Beets in Orange Sauce

12 medium beets
¾ teaspoon salt
SAUCE:
½ cup sugar
1½ tablespoons cornstarch
½ teaspoon salt
¼ teaspoon ground ginger
⅛ teaspoon pepper
⅓ cup wine vinegar
⅔ cup orange juice
2 tablespoons lemon juice
2 tablespoons butter

1. Remove tops from beets, leaving 1 inch of stem and all of root. Wash thoroughly but do not peel. Place beets in a large saucepan. Add cold water to cover and the ¾ teaspoon salt. Cover tightly, bring to a boil, reduce heat, and boil gently until tender, about 30 to 45 minutes. Drain thoroughly. Slip off skins and slice beets.

2. In a small saucepan, mix together sugar, cornstarch, the remaining ½ teaspoon salt, ginger, and pepper. Stir in vinegar. Cook over moderate heat, stirring constantly, until thickened and clear. Stir in orange and lemon juices and butter.

3. Add beets to sauce and cook over low heat until beets are hot. Serve immediately.

Yield: 4 to 6 servings

Lemon-Butter Broccoli

1 large bunch broccoli (about 2 pounds)
½ teaspoon salt
⅓ cup butter
2 tablespoons lemon juice

1. Wash broccoli, trim off bottom of stem, and remove any tough leaves. Cut stalks lengthwise into 6 pieces of about equal size.

2. In a skillet or large-bottomed saucepan bring to a boil approximately 1 inch water to which salt has been added. Add broccoli and boil, uncovered, 5 minutes. Cover tightly, reduce heat, and boil gently about 5 to 10 minutes longer or until crisp-tender. Drain thoroughly.

3. While broccoli is cooking, melt butter in a small saucepan. Stir in lemon juice.

4. Spoon butter sauce over cooked broccoli and serve immediately.

Yield: 6 servings

Brussels Sprouts with Mushrooms

1 quart Brussels sprouts
½ teaspoon salt
¼ cup butter
2 cups sliced mushrooms
½ teaspoon salt
⅛ teaspoon pepper
1 tablespoon lemon juice

1. Wash Brussels sprouts thoroughly and remove any damaged leaves. With a sharp knife cut a small cross in stem ends to allow for even cooking.

2. In a saucepan, bring to a boil approximately 1 inch water to which ½ teaspoon salt has been added. Add sprouts and boil, uncovered, 5 minutes. Cover tightly, reduce heat, and boil gently about 10 to 15 minutes longer or until tender. Drain thoroughly.

3. While sprouts are cooking, melt butter in a skillet. Add mushrooms and sauté until lightly browned and tender. Blend in remaining ½ teaspoon salt, pepper, and lemon juice.

4. Add mushroom mixture to cooked sprouts and toss gently to blend. Serve immediately.

Yield: 4 to 6 servings

Stuffed Cabbage with Tomato Sauce

1 large head green cabbage
1½ pounds lean ground beef
½ cup raw rice
½ cup chopped parsley
1 egg
1 teaspoon salt
⅛ teaspoon pepper
SAUCE:
3 tablespoons olive oil
3 large onions, coarsely chopped
2 cans (1 pound each) tomatoes OR
1 quart fresh tomatoes, cored and peeled
1 can (6 ounces) tomato paste
½ cup dark brown sugar, firmly packed
⅓ cup wine vinegar
1 bay leaf
1 teaspoon dill seed
¼ teaspoon ground allspice

1. Wash cabbage and remove any tough leaves. Bring to a boil enough water to cover cabbage head. Add cabbage and cook about 5 minutes, or until leaves are wilted and can be easily separated from the head. Drain thoroughly.

2. Remove cabbage leaves from head and cut out large center ribs with a sharp paring knife. Dry leaves with paper toweling. If inside leaves are too small for stuffing, leave them on head and cut head in wedges.

3. In a large bowl combine ground beef, rice, parsley, egg, salt, and pepper. Place a small amount of mixture in center of each leaf. Loosely fold sides of leaf over stuffing and roll up from end to make small rolls. Reserve any extra stuffing.

4. To make tomato sauce: In a large saucepan heat oil. Add onions and cook over moderate heat until limp and transparent. Add tomatoes, tomato paste, brown sugar, vinegar, bay leaf, dill seed, and allspice. Bring to a boil, stirring occasionally.

5. Into a large heavy pot spoon enough tomato sauce to cover bottom. Arrange a layer of cabbage rolls, seam side down, over sauce. Cover with a few spoonfuls of tomato sauce. Continue adding cabbage and sauce in layers, ending with sauce. If small head of inside leaves was cut in wedges, add to pot with cabbage rolls. Shape any extra stuffing in balls and add to pot.

6. Cover and cook over very low heat, so mixture barely simmers, 2 hours.

Yield: 4 to 6 servings

Cole Slaw with Fresh Dill

½ cup sour cream
½ cup mayonnaise
¼ cup wine vinegar
1 tablespoon sugar
2 teaspoons salt
¼ teaspoon pepper
3 quarts shredded or chopped cabbage (2 or 3 heads)
4 large sprigs dill, snipped

1. Cut cabbage in half and shred into fine slivers, or cut into quarters and chop finely.

2. In a small bowl blend together sour cream, mayonnaise, vinegar, sugar, salt, and pepper. Refrigerate.

3. Immediately before serving, add sour cream mixture to cabbage (which should be shredded just before use, if possible). Add dill and toss lightly to blend.

Yield: 6 to 8 servings

Baked Honey Carrots

2 tablespoons butter
½ teaspoon salt
⅛ teaspoon ground allspice
⅛ teaspoon pepper
2½ cups thinly sliced carrots (about 1 pound)
3 tablespoons honey

1. In a small ovenproof casserole, melt butter. Blend in salt, allspice, and pepper.

2. Add carrots and honey. Mix lightly to blend.

3. Cover tightly and bake in a preheated moderate oven (375 degrees) 40 minutes, or until carrots are tender. Mix lightly to blend and serve immediately.

Yield: 4 servings

Carrots in Broth

1½ pounds carrots
3 tablespoons butter
1 teaspoon sugar
⅛ teaspoon pepper
½ cup chicken broth
3 tablespoons finely chopped parsley

1. Wash and scrape carrots. Cut lengthwise in quarters and then into 2-inch-long strips. There should be about 5 cups.

2. In a large saucepan, melt butter. Blend in sugar, pepper, and chicken broth. Add carrots and toss gently to blend. Cover, bring to a boil, reduce heat, and boil gently about 10 to 15 minutes or until carrots are crisp-tender. Add parsley and toss gently to blend. Serve immediately.

Yield: 6 servings

Cauliflower with Cheese Sauce

½ teaspoon salt
1 large head cauliflower, separated into clusters

SAUCE:

2 tablespoons butter
2 tablespoons flour
¼ teaspoon salt
⅛ teaspoon pepper
4 drops Tabasco sauce
1 cup hot milk
¾ cup grated sharp Cheddar cheese
2 tablespoons snipped chives

1. In a saucepan, bring to a boil approximately 1 inch water to which the ½ teaspoon salt has been added. Add cauliflower and cover tightly. Bring to a boil, reduce heat, and boil gently about 8 to 15 minutes or until crisp-tender. Drain thoroughly.

2. While cauliflower is cooking, make sauce. In a small saucepan melt butter. Using a wire whisk, blend in flour. Season with the remaining ¼ teaspoon salt, pepper, and Tabasco.

3. Add milk to sauce and cook over moderate heat, stirring constantly, until thickened and smooth. Boil and stir 2 minutes. Blend in cheese.

4. Pour sauce over hot cooked cauliflower and garnish with chives. Serve immediately.

Yield: 6 servings

Celery and Potato Soup

3 tablespoons butter
2½ cups diced celery, including leaves
2 tablespoons flour
4 cups diced raw potatoes
6 cups hot water
2 teaspoons salt
¼ teaspoon pepper

1. In a large pot, melt butter. Add celery and cook over low heat, stirring occasionally, until glazed. Blend in flour.

2. Add potatoes, water, salt, and pepper. Simmer, partially covered, 40 minutes or until vegetables are tender. Stir occasionally and serve hot.

Yield: about 9 cups

Corn Cheese Custard

3 tablespoons butter
3 tablespoons flour
1 teaspoon salt
⅛ teaspoon pepper
1½ cups hot milk
½ cup grated sharp Cheddar cheese
2 cups corn (about 8 ears), cut off the cob
¼ cup thinly sliced scallions, including green tops
3 eggs, lightly beaten
¾ cup soft bread crumbs
3 tablespoons butter

1. In a saucepan melt 3 tablespoons butter. Using a wire whisk, blend in flour. Season with salt and pepper.

2. Add milk and cook over moderate heat, stirring constantly, until thickened and smooth. Boil and stir 2 minutes. Remove from heat and blend in cheese. Stir in corn and scallions.

3. Add a small amount of corn mixture to beaten eggs. Mix well. Add all of egg to corn mixture and blend well. Turn into a buttered 6-cup baking dish.

4. Sprinkle with crumbs and dot with remaining 3 tablespoons butter. Place dish in a shallow pan of hot water and bake in a preheated moderate oven (350 degrees) 55 to 60 minutes or until custard is set. Serve immediately.

Yield: 4 to 6 servings

Sweet Cucumber Pickles

4 quarts unpeeled cucumber slices, ¼ inch thick
2 cups boiling water
½ cup salt
1 quart white vinegar
3 cups sugar
2 tablespoons mustard seed
1½ teaspoons dill seed
½ teaspoon whole cloves

1. Place cucumber slices in a large bowl. Combine boiling water and salt, pour over cucumbers, and mix gently. Let stand overnight.

2. The next day, drain cucumbers and rinse well under cold water. Drain thoroughly.

3. In a large pot, combine vinegar, sugar, mustard seed, dill seed, and cloves. Bring to a boil, stirring occasionally. Add drained cucumbers and heat just until mixture comes to a boil. Remove from heat and pack into jars.

4. Seal jars and process in a boiling water bath (212 degrees) 5 minutes. (See canning directions on page 207).

Yield: about 6 pints

Eggplant Parmigiana

1 large eggplant (about 1 pound)
3 tablespoons flour, about
1 egg, beaten with 2 tablespoons water
½ cup dry bread crumbs, about
3 tablespoons olive oil, about
1 quart **tomato sauce** (page 184)
½ pound mozzarella cheese, sliced
⅓ cup grated Parmesan cheese

1. Wash eggplant. Cut off and discard a thin slice from stem end. Cut eggplant into crosswise slices about ¼ inch thick.

2. Dredge eggplant slices in flour. Dip in egg and then coat with bread crumbs. Refrigerate at least ½ hour.

3. In a large heavy skillet heat oil. Add eggplant, a few slices at a time, and cook over moderate heat until browned on both sides. Add more oil if necessary.

4. Pour about 1 cup tomato sauce into a 2½- to 3-quart shallow baking dish or casserole. Arrange a layer of eggplant on sauce. Cover with more sauce, then a layer of mozzarella slices and Parmesan cheese. Repeat with rest of eggplant, sauce, and cheeses.

5. Bake, uncovered, in a preheated moderate oven (350 degrees) 1 hour.

Yield: 4 to 6 servings

Eggplant with Sour Cream

1 large eggplant (about 1 pound)
2 tablespoons olive oil
⅔ cup coarsely chopped onion
½ teaspoon salt
⅛ teaspoon pepper
⅓ cup sour cream
2 teaspoons lemon juice
1 tablespoon snipped fresh dill
OR ½ teaspoon dried dill seed

1. Wash eggplant. Cut off and discard a thin slice from stem end. Cut eggplant into crosswise slices about ½ inch thick. Cut slices in cubes.

2. In a saucepan, bring approximately ½ inch water to a boil. Add eggplant and cover tightly. Bring to a boil, reduce heat, and boil gently about 5 to 7 minutes or until tender. Drain thoroughly.

3. In another saucepan heat oil. Add onion and cook over moderate heat, stirring occasionally, until golden. Blend in salt and pepper.

4. Add cooked eggplant and sour cream. Cook and stir gently over low heat until mixture is hot. Do not boil. Immediately before serving stir in lemon juice and dill.

Yield: 4 servings

Leeks Vinaigrette

12 leeks, about 1 inch in diameter
½ teaspoon salt
½ cup olive oil
3 tablespoons wine vinegar
¼ teaspoon salt
⅛ teaspoon pepper
2 teaspoons capers
2 to 3 tablespoons finely chopped parsley

1. With a sharp knife, cut off roots and tops of leeks, leaving about 2 inches of green leaves. Cut leeks lengthwise to within 1 inch of root end. Wash thoroughly under cool running water to remove sand.

2. In a saucepan, bring to a boil about 1 inch water to which the ½ teaspoon salt has been added. Add leeks and boil, uncovered, 5 minutes. Cover tightly, reduce heat, and boil gently 2 to 5 minutes longer or until tender. Remove leeks with a slotted spoon or pancake turner, drain thoroughly, and arrange in a shallow dish. Cool to room temperature.

3. Meanwhile combine oil, vinegar, remaining ¼ teaspoon salt, pepper, and capers. Stir vigorously to blend. Pour over leeks and refrigerate until chilled, at least 1 hour. Immediately before serving, garnish with chopped parsley.

Yield: 4 to 6 servings

Classic Vichyssoise

¼ cup butter
3 cups thinly sliced leeks, white part only
4 medium potatoes (about 1⅓ pounds), peeled and thinly sliced
5 cups chicken broth
1 cup milk
1 cup heavy cream
Snipped fresh chives

1. In a large pot or kettle melt butter. Add leeks and cook over low heat, stirring occasionally, until very limp and tender. Do not brown.

2. Add potatoes and broth. Bring to a boil, reduce heat, and simmer, partially covered, 35 to 40 minutes or until vegetables are very tender. Stir in milk.

3. Remove from heat. Purée in an electric blender, or crush and put through a fine sieve. Stir in cream and refrigerate until thoroughly chilled. Though usually served cold, vichyssoise may also be served hot. Before serving, garnish each bowlful with snipped chives.

Yield: about 9 cups

Lettuce Salad with Anchovy Dressing

Lettuce is best eaten plain with a tangy oil-and-vinegar dressing.

3 anchovy fillets OR ¼ teaspoon anchovy paste
½ teaspoon prepared mustard
¼ cup olive oil
2 tablespoons wine vinegar
1/16 teaspoon black pepper
Dash garlic powder

1. In a small bowl, mash anchovies to a paste. Add mustard and mix well.
2. Add oil, vinegar, pepper, and garlic powder, and mix vigorously to blend. Serve over lettuce or other green salads.

Yield: scant ½ cup

Peas with Celery and Scallions

½ teaspoon salt
2 cups shelled peas
3 tablespoons butter
⅛ teaspoon pepper
3 medium scallions (including green tops), cut diagonally in 1-inch pieces
½ cup thinly sliced celery

1. In a saucepan, bring to a boil approximately 1 inch water to which salt has been added. Add peas and boil, uncovered, about 5 minutes. If peas are young, they may be done at this point. If not, cover tightly, reduce heat, and boil gently 3 to 5 minutes longer or until tender. Drain thoroughly and return to pot.
2. While peas are cooking, melt butter in a small skillet. Blend in pepper. Add scallions and celery, and cook over low heat, stirring occasionally, until vegetables are glazed and celery is tender.
3. Add scallion-celery mixture to cooked peas and toss gently to blend. Serve immediately.

Yield: 4 servings

Minted Peas in Cream

Fresh carrots may be substituted for the peas. See directions below.

3 tablespoons butter
2 teaspoons sugar
¼ teaspoon salt
⅛ teaspoon pepper
2 cups shelled peas
1 teaspoon flour
1 teaspoon fresh snipped mint OR ½ teaspoon dried mint
½ cup heavy cream

1. In a medium skillet, melt butter. Blend in sugar, salt, and pepper.

2. Add peas and toss lightly to blend. Cover and simmer gently, stirring occasionally, about 5 minutes or until peas are tender.

3. Blend in flour. Add mint and cream. Cook, stirring gently, until mixture comes to a boil and thickens slightly. Serve immediately.

VARIATION: Substitute 2½ cups thinly sliced carrots for peas. Increase cooking time to 10 minutes, or until carrots are crisp-tender.

Yield: 4 servings

Stuffed Green Peppers

4 large (or 6 medium) green peppers
2 tablespoons olive oil
¾ cup finely chopped onion
1 pound lean ground beef
1½ cups cooked rice
3 tablespoons chopped parsley
¾ teaspoon salt
1 teaspoon thyme leaves
OR ½ teaspoon dried thyme
¼ teaspoon pepper
1 teaspoon Worcestershire sauce
2½ cups **tomato sauce** (page 184)

1. Wash peppers. Cut off and discard a thin slice from stem ends. Remove seeds and rib portion and discard.

2. In a large pot bring 1 cup water to a boil. Add peppers. Cover and boil 5 minutes. Drain thoroughly.

3. In a large skillet heat oil. Add onion and cook over moderate heat, stirring occasionally, until limp and transparent. Add ground beef and cook, stirring to break into small pieces, until browned. Blend in rice, parsley, salt, thyme, pepper, and Worcestershire sauce.

4. Stuff peppers with meat mixture. Arrange upright in a shallow baking pan or casserole and spoon tomato sauce around peppers. Bake in a preheated moderate oven (375 degrees) 30 minutes. Spoon sauce from pan over peppers before serving.

Yield: 4 to 6 servings

Easy Fried Potatoes

3 tablespoons salad oil, about
1 medium onion, thinly sliced
4 medium potatoes, peeled and sliced ¼ inch thick
½ teaspoon salt
⅛ teaspoon pepper

1. In a large heavy skillet, heat oil. Add onion and cook over moderate heat, stirring occasionally, until golden.

2. Add potatoes. Cover and cook, turning occasionally, until potatoes are tender and lightly browned. Add more oil if potatoes begin to stick to pan. Season with salt and pepper, and serve immediately.

Yield: 4 servings

Fried Potato Cakes

⅓ cup butter, about
½ cup finely chopped onion
3 cups cold homemade mashed potatoes
2 eggs
2 tablespoons finely chopped parsley

1. In a large heavy skillet or on a griddle, melt 1 tablespoon of the butter. Add onion and cook over low heat, stirring occasionally, until lightly browned.

2. Meanwhile, mix together mashed potatoes, eggs, and parsley. Add onion and mix well.

3. Melt remaining butter in a skillet or on a griddle. Form potato mixture into patties of about ¼ cup each. Spoon patties into skillet. Cook over moderate heat until brown on bottom. Turn carefully with wide metal spatula and brown the other side. Add more butter if necessary. Serve immediately.

Yield: 6 servings

Mashed Potato Soufflé

⅓ cup butter
¼ cup dry bread crumbs, about
2½ cups unseasoned hot homemade mashed potatoes
½ cup hot milk
¾ cup grated sharp Cheddar cheese
½ teaspoon salt
⅛ teaspoon pepper
2 eggs, separated

1. Using 1 tablespoon of the butter, grease a 6-cup soufflé dish. Sprinkle with crumbs and shake out extra crumbs.

2. In a bowl, mix together mashed potatoes, milk, cheese, salt, pepper, remaining butter, and egg yolks. Beat until well blended.

3. In a small bowl, beat egg whites until stiff. Do not overbeat.

4. Fold about one-third of beaten egg whites into the potato mixture. Then fold in remaining whites.

5. Turn mixture into the prepared dish and bake in a preheated moderate oven (375 degrees) 35 to 40 minutes, or until puffy and brown. Serve immediately.

Yield: 4 to 6 servings

New Potato Salad

½ teaspoon salt
2 pounds new potatoes, washed but not peeled
¾ cup mayonnaise
1 teaspoon salt
¼ teaspoon pepper
1 teaspoon prepared mustard
2 tablespoons wine vinegar
1½ cups diced celery
3 hard-cooked eggs

1. In a large saucepan, bring to a boil approximately 1 inch water to which the ½ teaspoon salt has been added. Add potatoes and cover tightly. Bring to a boil, reduce heat, and boil gently about 15 to 20 minutes or until potatoes are tender. Cooking time will depend on size of potatoes. Drain thoroughly. Peel and leave whole or cut in pieces.

2. While potatoes are cooking, in a small bowl combine mayonnaise, the remaining 1 teaspoon salt, pepper, mustard, and vinegar. Blend well.

3. In a large bowl, combine potatoes, celery, and mayonnaise dressing. Toss very gently to blend. Refrigerate until chilled.

4. Immediately before serving, cut eggs into thick slices and add to salad. Mix lightly to blend.

Yield: 6 to 8 servings

Ratatouille

Originating in Nice on the French Riviera, ratatouille is a marvelous combination of eggplant, tomatoes, zucchini, and green pepper. It may be served hot or cold.

½ cup olive oil
2 medium onions, thinly sliced
1 clove garlic, finely minced
1 medium green pepper, thinly sliced
2 medium zucchini, sliced
1 medium eggplant, unpeeled, cut in 1-inch cubes
5 medium tomatoes, cored, peeled, and coarsely chopped
3 tablespoons finely chopped parsley
2 teaspoons fresh snipped basil leaves OR 1 teaspoon dried basil
1 teaspoon salt
¼ teaspoon pepper
Grated Parmesan cheese (optional)

1. In a large skillet, heat about ¼ cup of the oil. Add onion, garlic, and green pepper and cook over low heat, stirring occasionally, until onion is limp and transparent. Remove to a small bowl and reserve.

2. Add remaining oil to skillet and add zucchini and eggplant. Cook over moderate heat, mixing gently, until vegetables are lightly glazed. Add reserved onion mixture and blend gently.

3. Add tomatoes. Sprinkle with parsley and basil and season with salt and pepper. Cover and simmer gently 20 minutes, stirring occasionally. Uncover and cook 10 minutes longer or until some of the liquid has evaporated. If desired, sprinkle with Parmesan cheese before serving. Serve hot or cold.

Yield: 6 or more servings

Spinach with Sour Cream and Horseradish

2 pounds spinach
½ cup sour cream
2 tablespoons prepared horseradish
¼ teaspoon salt
⅛ teaspoon pepper

1. Remove tough stems from spinach and pick out imperfect leaves. Wash several times in a large amount of warm water. After each washing, lift gently out of water to let sand and grit settle to bottom. Drain.

2. Place spinach in a large pot. Water that clings to the leaves will be sufficient. Do not add more water. Cover pot tightly and cook 3 to 10 minutes or until spinach is wilted and tender. Drain thoroughly and chop fine.

3. In a saucepan combine sour cream, horseradish, salt, and pepper. Add spinach and cook over low heat, mixing gently, until sauce is blended with spinach and mixture is hot. Do not allow to boil. Serve immediately.

Yield: 4 servings

Baked Acorn Squash

3 acorn squash
⅓ cup dark brown sugar, firmly packed
¾ teaspoon salt
¾ teaspoon ground cinnamon
¼ teaspoon ground nutmeg
¼ teaspoon pepper
3 tablespoons butter

1. Cut squash in half lengthwise and spoon out seeds and stringy center portion. Arrange squash, cut side up, in a shallow baking dish.

2. In a small bowl combine brown sugar, salt, cinnamon, nutmeg, and pepper. Spoon equal amounts of mixture into cavities of squash. Dot with butter.

3. Bake in a preheated moderate oven (350 degrees) 60 minutes, or until squash is tender. About 10 minutes before removing from oven, brush cut surfaces with some of the butter mixture to glaze. Serve immediately.

Yield: 6 servings

Mashed Butternut Squash

3 pounds butternut squash
½ teaspoon salt
¼ cup orange juice
1½ tablespoons butter
½ teaspoon salt
⅛ teaspoon pepper
⅛ teaspoon ground ginger

1. Wash squash. Cut off and discard a thin slice from stem and blossom ends. Cut squash into slices about 2 inches thick and then cut each slice into quarters. Peel and remove seeds and stringy center portion.

2. In a large saucepan, bring to a boil approximately 1 inch water to which ½ teaspoon salt has been added. Add squash and cover tightly. Bring to a boil, reduce heat, and boil gently until tender, about 15 to 25 minutes. Drain thoroughly.

3. Using a potato masher or electric mixer, mash squash until smooth. Add to squash the orange juice, butter, remaining ½ teaspoon salt, pepper, and ginger. Blend well. Serve immediately.

Yield: 6 servings

Herbed Summer Squash

¼ cup butter
⅓ cup thinly sliced scallions, including green tops
1 teaspoon thyme leaves OR ½ teaspoon dried thyme
½ teaspoon oregano leaves OR ¼ teaspoon dried oregano
¼ teaspoon salt
⅛ teaspoon pepper
1 quart summer squash (about 1 pound), thinly sliced
¼ cup finely chopped parsley

1. In a large skillet melt butter. Blend in scallions, thyme, oregano, salt, and pepper.

2. Add squash and mix lightly to blend.

3. Cover and cook over moderate heat, stirring occasionally, 10 to 15 minutes or until squash is tender. Sprinkle with parsley and serve immediately.

Yield: 3 to 4 servings

Baked Tomato Halves

6 medium tomatoes
¾ cup soft bread crumbs
½ cup grated sharp Cheddar cheese
2 teaspoons sugar
1 teaspoon salt
¾ teaspoon chili powder
¼ teaspoon pepper
¼ cup butter

1. Core tomatoes and cut in half crosswise. Arrange, cut side up, in a lightly greased, shallow baking dish.

2. In a small bowl, combine bread crumbs, cheese, sugar, salt, chili powder, and pepper. Spoon equal parts of mixture over tomatoes. Dot with butter.

3. Bake in a preheated moderate oven (350 degrees) 30 minutes, or until crumbs are lightly browned and tomatoes are tender. Serve immediately.

Yield: 6 servings

Tomato Sauce

Excellent over spaghetti, or use for **eggplant parmigiana** (page 176) or **stuffed green peppers** (page 179).

3 tablespoons olive oil, about
1½ cups coarsely chopped onion
1 cup chopped celery
1 cup chopped green pepper
2 quarts tomatoes, cored, peeled, and cut in quarters
1 can (6 ounces) tomato paste
1 tablespoon sugar
1 tablespoon salt
2 bay leaves
¼ teaspoon pepper
1½ teaspoons basil leaves OR ¾ teaspoon dried basil
1 teaspoon oregano leaves OR ½ teaspoon dried oregano

1. In a large saucepan or pot, heat oil. Add onion, celery, and green pepper, and cook over low heat, stirring occasionally, until onion is limp and transparent and vegetables are glazed. Add more oil if necessary.

2. Add tomatoes, tomato paste, sugar, salt, bay leaves, and pepper. Simmer slowly, uncovered, stirring occasionally, about 1½ hours or until thickened. Exact cooking time will depend on juiciness of tomatoes.

3. Add basil and oregano, and cook 10 minutes longer.

Yield: about 2 quarts

Mashed Turnips

1½ pounds turnips
½ teaspoon salt
2 tablespoons butter
2 tablespoons dark brown sugar
¼ teaspoon salt
¼ teaspoon ground ginger
⅛ teaspoon pepper
1 teaspoon lemon juice
4 tablespoons hot milk, about

1. Wash turnips and pare thinly. Cut in pieces of uniform size.

2. In a saucepan bring to a boil approximately 1 inch water to which ½ teaspoon salt has been added. Add turnips and cover tightly. Bring to a boil, reduce heat, and boil gently until tender, about 10 to 20 minutes, depending on size of pieces. Drain thoroughly.

3. Using a potato masher or electric mixer, mash turnips until fairly smooth. Add all remaining ingredients and blend well. If necessary, add more milk until mixture is about the consistency of mashed potatoes. This dish may be made in advance up to this point.

4. Turn into a buttered ovenproof casserole and bake, covered, in a moderate oven (350 degrees) until hot, about 35 minutes.

Yield: 4 to 6 servings

Zucchini and Mushroom Omelet

Serve this as the delicious entrée for a lunch or supper for two, or as one course in a dinner for four.

3 tablespoons butter
1 quart zucchini (about 1 pound), thinly sliced
1½ cups sliced mushrooms
4 eggs
¼ cup milk
¼ teaspoon salt
⅛ teaspoon pepper
½ cup grated Parmesan cheese

1. In a 10-inch skillet with ovenproof handle, melt butter. Add zucchini and mix lightly to coat with butter.

2. Cover and cook over moderate heat, stirring occasionally, about 10 minutes or until almost tender.

3. Add mushrooms, cover, and cook, stirring once or twice, about 5 minutes longer or until mushrooms are lightly browned and tender.

4. Meanwhile, beat eggs lightly with milk, salt, and pepper. Pour over zucchini-mushroom mixture and bake in a preheated moderate oven (350 degrees) 20 minutes, or until eggs are set.

5. Sprinkle with cheese and place under broiler until cheese melts. Serve immediately.

Yield: 2 to 4 servings

DESSERTS

Blueberry Crumb Cake

¼ cup butter
1 teaspoon vanilla extract
½ cup sugar
1 egg
1 cup flour
1 teaspoon baking powder
¼ teaspoon salt
⅓ cup milk
1½ cups blueberries, washed and drained thoroughly
Crumb Topping

Vanilla or maple walnut ice cream (*optional*)

1. In a mixing bowl, cream butter until light. Blend in vanilla. Add sugar and beat until light and fluffy. Beat in egg.

2. Sift together flour, baking powder, and salt. Add to creamed mixture alternately with milk, starting and ending with dry ingredients. Stir well after each addition. Turn into a greased 9-inch-square baking pan. Batter will just cover bottom of pan.

3. Pour blueberries on top of batter. Sprinkle with **Crumb Topping** and bake in a preheated moderate oven (375 degrees) 40 to 45 minutes, or until a cake tester inserted in the center comes out clean. Serve warm, plain or with ice cream as desired.

Crumb Topping

½ cup sugar
⅓ cup flour
¼ cup butter, at room temperature

In a medium bowl, combine sugar and flour. Add butter and cut in with a pastry blender until mixture is crumbly and about the consistency of coarse cornmeal.

Yield: 9-inch cake

Blueberry Sauce

This sauce may be served over cakes, custards, ice cream, or puddings.

¾ cup water
¼ cup sugar
2 teaspoons cornstarch
1 tablespoon cold water
1½ cups blueberries, washed and drained

1. In a saucepan, mix the ¾ cup water and sugar. Bring to a boil, stirring frequently.

2. In a small custard cup, combine cornstarch and cold water. Mix until cornstarch is dissolved. Add to sugar syrup and boil, stirring constantly, until thickened and smooth.

3. Add blueberries, bring to a boil, and cook 2 minutes, stirring gently. Serve warm or cold over the dessert.

Yield: about 1⅔ cups

Blueberry Turnovers

FILLING
½ cup sugar
¼ teaspoon salt
¼ cup cornstarch
⅓ cup cold water
2 cups blueberries, washed and drained thoroughly

PASTRY
2 cups flour
¼ teaspoon salt
3 tablespoons sugar
⅔ cup shortening
½ cup cold water, about
1 egg white

Vanilla or coffee ice cream (*optional*)

1. Make filling: In a saucepan, mix together the ½ cup sugar, ¼ teaspoon salt, and cornstarch. Add cold water and cook over moderate heat, stirring constantly, until mixture comes to a boil. Boil and stir one minute. Remove from heat and add berries.

2. Stir gently, trying to keep berries whole, until berries are coated with sugar mixture. Refrigerate while making pastry.

3. Make pastry: in a mixing bowl, combine flour, ¼ teaspoon salt and the 3 tablespoons sugar. Using a pastry blender, cut in shortening until mixture is crumbly and the texture varies from about the size of small peas to that of coarse cornmeal.

4. Add the ½ cup cold water to pastry, about a tablespoon at a time, stirring it in with a fork, until mixture holds together to form a dough.

5. Divide dough into eight pieces. On a lightly floured surface roll each piece into a circle about 6½ inches in diameter. Add more flour if dough is sticky and becomes difficult to work with. Circles need not be perfect, because turnovers can be shaped after they are filled.

6. As circles are rolled out, arrange them on two large cookie sheets. Spoon about ¼ cup blueberry filling on half of each circle; fold over pastry to form a turnover and seal edges. Refrigerate turnovers at least ½ hour.

7. Prick turnovers with a fork in about 3 places to allow steam to escape. Beat egg white until foamy and use to brush turnovers. Bake in a preheated hot oven (425 degrees) 15 to 20 minutes or until lightly browned. Serve warm with ice cream or plain cream.

Yield: 8 turnovers

Pumpkin Spice Cake with Lemon Frosting

1 cup pumpkin purée
1 cup sugar
¾ teaspoon ground cinnamon
½ teaspoon ground ginger
¼ teaspoon ground nutmeg
⅛ teaspoon ground cloves
⅔ cup vegetable oil
2 eggs
1½ cups sifted flour
1 teaspoon baking powder
1 teaspoon baking soda
½ teaspoon salt

1. Make purée by putting cooked pumpkin through a sieve or food mill.

2. In a mixing bowl, combine the sugar, cinnamon, ginger, nutmeg, and cloves.

3. Add pumpkin and oil and beat until smooth. Add eggs, one at a time, beating well after each addition.

4. Sift together flour, baking powder, soda, and salt. Stir into batter.

5. Pour into a greased 9 x 5 x 3-inch loaf pan and bake in a preheated moderate oven (350 degrees) 55 to 60 minutes, or until cake springs back when lightly touched in the center. Place pan on rack and cool for 10 minutes. Remove from pan and finish cooling. Frost top with **Lemon Frosting.**

Lemon Frosting

3 tablespoons butter, at room temperature
1½ cups sifted confectioners' sugar
1 tablespoon lemon juice

Combine all ingredients and beat vigorously until smooth. If necessary, add a small amount of milk to bring to spreading consistency.

Yield: 1 loaf cake

Raspberry Pie

PASTRY
2 cups flour
¼ teaspoon salt
2 tablespoons sugar
⅔ cup shortening
½ cup cold water, about

FILLING
¾ cup sugar
2 tablespoons flour
⅛ teaspoon ground nutmeg
⅛ teaspoon salt
1 quart raspberries
2 tablespoons butter

Vanilla or butter pecan ice cream (*optional*)

1. Make pastry: In a mixing bowl combine the 2 cups flour, ¼ teaspoon salt and 2 tablespoons sugar. Using a pastry blender, cut in shortening until mixture is crumbly and the texture varies from about the size of small peas to that of coarse cornmeal.

2. Add cold water, about 1 tablespoon at a time, stirring it in with a fork, until mixture holds together to form a dough.

3. Divide dough in half. On a lightly floured surface, roll one half into a circle about 12 inches in diameter. Add more flour if dough is sticky and becomes difficult to work with. Fold pastry in half and arrange in a 9-inch pie plate with fold in center of plate. Unfold and gently line pie pan with pastry.

4. Make filling: In a large bowl mix together the ¾ cup sugar, flour, nutmeg, and salt. Add berries and toss gently to coat with sugar mixture. Try to keep berries whole. Spoon berries and sugar mixture into prepared pastry in pie pan. Dot with butter.

5. Roll out remaining pastry into a circle about 12 inches in diameter. Fold in half, place over berry filling, and unfold. Seal edges and cut several slits in the top crust for steam to escape. Bake in a preheated hot oven (400 degrees) 40 to 50 minutes, or until crust is browned and berries are tender. Serve warm, with ice cream or plain.

Yield: 9-inch pie

Rhubarb Crisp

1 cup sugar
⅓ cup flour
2 quarts rhubarb (about 2½ pounds), cut in 1½-inch pieces
1 cup flour
½ cup uncooked oatmeal
½ cup dark brown sugar, firmly packed
½ cup sugar
½ cup butter, melted and cooled to room temperature
Butter pecan or coffee ice cream (*optional*)

1. In a large bowl combine the 1 cup sugar and ⅓ cup flour. Add rhubarb and toss lightly until rhubarb is coated with sugar mixture. Spoon into 13 x 9 x 2-inch baking dish.

2. In the same bowl combine the 1 cup flour, oatmeal, brown sugar, and remaining ½ cup sugar. Add butter and cut in with pastry blender until mixture is crumbly.

3. Sprinkle crumbs over rhubarb. Bake in a preheated moderate oven (375 degrees) 35 minutes, or until topping is brown and rhubarb is tender. Serve warm, with ice cream if desired.

Yield: 8 to 10 servings

Rhubarb Shortcake

2 cups flour
3 tablespoons sugar
1 tablespoon baking powder
½ cup shortening
1 egg, lightly beaten
7 tablespoons milk, about
Butter
Stewed Rhubarb (below)
1½ cups heavy cream, plain or whipped

1. In a mixing bowl, sift together flour, sugar, and baking powder. Using a pastry blender, cut in shortening until mixture is crumbly and resembles coarse cornmeal.

2. Add beaten egg and milk, about a tablespoon at a time, stirring until mixture holds together to form a soft dough. Turn out onto floured surface and knead 10 times.

3. Roll out to a circle about ¾ inch thick. Cut in 2½-inch rounds with a biscuit cutter or glass that has been dipped in flour. Arrange biscuits on cookie sheet and bake in a preheated hot oven (400 degrees) 15 minutes or until golden brown.

4. While biscuits are still warm, split in half and generously butter cut sides. To serve: Place bottom of biscuit in individual dessert bowl. Top with some of stewed rhubarb. Cover with remaining biscuit half and more rhubarb. Top with cream, either plain or whipped.

Yield: 8 servings

Stewed Rhubarb

Use for **Rhubarb Shortcake** (above) or serve plain as a fruit dessert.

1 cup water
1¼ cups sugar
2 quarts rhubarb (about 2½ pounds), cut in 1½-inch pieces

1. In a medium-size saucepan, combine water and sugar. Bring to a boil, stirring frequently.

2. Add rhubarb and bring to a boil. Reduce heat and simmer gently, covered, about 10 minutes or until tender. Serve warm or chilled.

Yield: 8 to 10 servings, or about 5½ cups

Squash Pie

1⅔ cups butternut squash purée
3 eggs
¾ cup sugar
1½ teaspoons ground cinnamon
1 teaspoon ground ginger
½ teaspoon ground nutmeg
¼ teaspoon ground cloves
¼ teaspoon salt
1 can (13¾ ounces) evaporated milk, undiluted
1 9-inch pie shell, unbaked
Whipped cream (*optional*)

1. Make purée by putting cooked squash through a sieve or food mill.

2. In a large mixing bowl, beat eggs until frothy. Blend in sugar, cinnamon, ginger, nutmeg, cloves, and salt.

3. Add squash and milk; mix until smooth.

4. Pour into pie shell and bake in a preheated hot oven (400 degrees) 55 to 60 minutes, or until a knife inserted in the center comes out clean. Serve with whipped cream if desired.

Yield: 9-inch pie

Mike's Strawberries-and-Cream Cake

½ cup butter
1 teaspoon vanilla extract
1 cup sugar
3 eggs, separated
1¾ cups flour
2 teaspoons baking powder
½ teaspoon salt
½ cup milk
Strawberry Sauce
1 cup heavy cream, whipped

1. In a mixing bowl, cream butter until light. Blend in vanilla. Add sugar and beat until light and fluffy. Add egg yolks, one at a time, beating well after each addition.

2. Sift together flour, baking powder, and salt. Add to creamed mixture alternately with milk, starting and ending with dry ingredients. Stir well after each addition.

3. Beat egg whites until stiff but not dry and fold into batter. Turn into a greased 9 x 5 x 3-inch loaf pan and bake in a preheated moderate oven (350 degrees) 45 to 55 minutes, or until cake springs back when lightly touched in the center. Cool on rack for 10 minutes. Remove from pan and finish cooling.

4. To serve: Cut cake in slices about 1 inch thick. Spoon **Strawberry Sauce** over cake and top with whipped cream.

Strawberry Sauce

1½ quarts strawberries
¾ cup sugar

Wash and hull berries. Place in a large shallow pan and crush with a potato masher. Sprinkle with sugar and refrigerate until chilled and some of the juice is drawn out of the berries, at least an hour.

Yield: 8 servings

Strawberry Cheesecake

CRUST
1 cup flour
½ cup sugar
1 teaspoon baking powder
⅛ teaspoon salt
½ cup butter, at room temperature
1 egg

FILLING
1 package (8 ounces) cream cheese, at room temperature
1 teaspoon vanilla extract
½ teaspoon almond extract
⅔ cup sugar
3 eggs
1 cup sour cream
3 cups strawberries

GLAZE
2 cups strawberries
½ cup sugar
4 teaspoons cornstarch

1. Make crust: In a mixing bowl, combine flour, sugar, baking powder, and salt. Add butter and cut in with pastry blender until mixture is crumbly. Add egg and mix to form a dough.

2. Using fingertips, press dough on bottom and halfway up sides of 9-inch-square baking pan. If mixture sticks to hands, dip fingers in flour. Refrigerate crust while making filling.

3. Make filling: In a mixing bowl, beat cream cheese until light. Blend in vanilla and almond extracts. Add sugar and beat until light and fluffy. Add eggs, one at a time, beating well after each addition. Stir in sour cream. Turn into prepared crust and bake in a preheated moderate oven (350 degrees) 35 minutes, or until crust is lightly browned and custard is set. Place pan on rack until cake is cool and then refrigerate until chilled.

4. Make glaze: Wash and hull the 2 cups of strawberries. Place in a medium saucepan and crush with a potato masher. Combine the ½ cup sugar and cornstarch, and add to berries. Cook over moderate heat, stirring constantly, until mixture comes to a boil and thickens. Pour into container of electric blender and blend until smooth. Cool thoroughly.

5. Wash, hull, and thoroughly drain the remaining 3 cups of strawberries. Arrange, hulled side down, on cake. Spoon cooled glaze over berries and refrigerate at least 2 hours before serving.

Yield: 9-inch cake

Strawberry Ice Cream Sauce

3 cups strawberries
½ cup sugar

1. Wash and hull strawberries. Place 2 cups of the berries in container of electric blender with the sugar. Blend quickly on low speed to combine ingredients. If possible, remove from blender while there are still some pieces of berries in sauce.

2. Slice remaining 1 cup berries and stir into sauce. Refrigerate before serving. Serve over ice cream.

Yield: about 2⅓ cups

11.

Storing, Freezing, and Canning

Even if summer is gone and the dark days of winter are upon you, the vegetables you tended so carefully need not disappear from your table. You can still savor them and benefit both your table and your budget. And you can enjoy succulent varieties that you would never find in the market in winter—and probably not in summer. There are three ways to achieve this: the time-honored ways of storing in the cellar or outdoors and of canning, and the modern way of preserving vegetables by freezing them. Each is good and each needs certain procedures in order to be successful.

STORING VEGETABLES INDOORS OR OUTDOORS

If kept under the proper conditions, vegetables will retain their garden-fresh flavor long after harvest and brighten many a winter meal. The main requirements are low temperature, moisture, and darkness, all of which are easily provided by sectioning off part of the cellar or crawl space. If you have no cellar, you can store outdoors. Remember:

1. Vegetables must be stored in darkness.
2. Temperatures are best when hovering around 32 degrees.
3. Most crops will shrivel if not given high humidity (around 85 percent). For exceptions, see the chart on page 195.
4. Stored vegetables need good ventilation.

Cellar Storage A well-ventilated basement with or without central heating is an ideal location for a storage closet. If the furnace is nearby, however, insulating partitions should be built. Make provision for ventilation by locating the storage room near a window. Then construct walls of two-by-fours and siding material—either boards or plywood. For protection against rats and mice, line the walls and ceiling with one-fourth-inch wire mesh before nailing on the wall covering. Shelves and bins can be constructed from clean leftover scrap materials from other projects. If your cellar has a dirt floor this is all the better for maintaining high humidity. A cement floor should be sprinkled with water occasionally to provide humidity.

Outdoor Storage If you have no cellar or the cellar is earmarked for other purposes, hardier vegetables—potatoes, carrots, beets, turnips, parsnips, and cabbage—can be stored outdoors. Outdoor storage should be attempted only in regions where the temperature stays between 28 and 55 degrees in winter. To make certain outdoor storage is practiced successfully in your region, consult the county agent.

Outdoor storage is rather impractical and old-fashioned, though it works and for some is the only method possible. First spread a layer of straw or other bedding material such as leaves and soil on the ground or in a shallow pit. Then stack individual vegetables in cone-shaped piles, putting more straw between each layer of vegetables. Cover the piles with straw, then with four to six inches of soil, making sure the soil does not come into direct contact with the vegetables. Tamp the soil with a shovel to make the piles as waterproof as possible.

After a pile has been opened in the winter, it is impossible to close it up again; therefore all vegetables should be removed at that time. To prolong the pleasures of fresh vegetables in the winter, plan on several small piles. A pile of carrots, for example, should provide no more than a two-week supply.

Handling Vegetables for Storage Vegetables slated for the storage bin should be as free of bruises, cuts, and imperfections as possible in order to keep down losses from disease and rotting. To get the most out of stored vegetables:

Gently rinse remaining soil from vegetables.

Handle produce carefully to avoid scratching, gouging, or nicking the skin or rind.

Use containers with smooth inner surfaces free of protruding staples, nails, or splinters that can dig into vegetables.

Check stored produce regularly and remove immediately any vegetables that appear to be decaying.

When preparing for storage, leave a few inches of stems or tops attached.

After picking, allow rind crops such as winter squash and pumpkin to remain in the field several days to harden skins.

Not all vegetables need the cool moist conditions outlined above. Soft vegetables—ripe tomatoes and greens—cannot be stored for long. Other exceptions are pumpkins and squash, which need a warm dry location, and onions, dry peas, and beans, which store best under cool dry conditions. Refer to the following chart before putting vegetables away for the winter. (See chapter 8 for directions for drying and storing herbs for later use.)

FREEZING FRESH FRUITS AND VEGETABLES

The joy of fresh fruits and vegetables no longer ends with the first frost of fall, as in the days when there were no home freezers. Now the home freezer stocks everything from ice cream sandwiches to chicken tetrazzini and beets in orange sauce. It's terrific for families who garden. As the vegetables are harvested, they can be frozen

IDEAL LONG-TERM STORAGE CONDITIONS FOR INDIVIDUAL CROPS*

VEGETABLE	FREEZING POINT	WHERE TO STORE	TEMPERATURE	HUMIDITY	WHEN TO USE
Dry beans and peas	—	Any cool dry place	32° to 40°	Dry	Whenever desired
Fall cabbage	30.4	Outdoor pit	32°	Moderately moist	Late fall and winter
Cauliflower	30.3	Cellar storage bin	32°	Moderately moist	Within 6 to 8 weeks
Fall celery	31.6	Pit or storage cellar	32°	Moderately moist	Late fall and winter
Onions	30.6	Any cool dry place	32°	Dry	Through fall and winter
Parsnips	30.4	Cellar storage bin	32°	Moist	Through fall and winter
Peppers	30.7	Unheated basement	45° to 50°	Moderately moist	Within 2 to 3 weeks
Potatoes	30.9	Cellar storage bin	35° to 40°	Moderately moist	Through fall and winter
Pumpkins and winter squashes	30.5	Cellar or basement	55°	Moderately dry	Through fall and winter
Root crops (beets and carrots)	—	Cellar storage bin	32°	Moist	Through fall and winter
Tomatoes	31.0	Cellar storage bin	55° to 70°	Moderately dry	Within 4 to 6 weeks

*Adapted from *Home and Garden Bulletin No. 119*, U.S. Department of Agriculture.

either plain or according to a special recipe and stacked in the freezer. There they sit, most conveniently, until minutes before you are ready to serve them.

Choosing the Right Containers Once you have decided to do some freezing, the first step is to get the right containers. There are lots of kinds to pick from and the choice depends on what's available, how much money you want to spend, and what you plan to freeze.

To qualify for freezing, any container must be able to protect food from losing moisture. It should be durable enough to survive at freezing temperatures and it must seal tightly, especially if you are going to pack anything in syrup. Another point to keep in mind is the number of servings you expect to get from each container. Buy a size that will hold enough for one family-size portion.

The next consideration is the shape. Square containers with flat tops and bottoms stack well and don't waste freezer space. Round containers and those with flared sides also stack nicely but there is the problem of wasted space. Polyethylene bags are inexpensive and can hold a lot of food, but they have absolutely no shape and bulge all over. Again, wasted space is inevitable.

Take a good look at the design of the container. If the mouth is smaller than the body, it will be impossible to remove the food before it thaws. Glass jars often have this troublesome shape and present a terrible problem when you are in a hurry. One last tip: if the empty containers can be folded or stacked inside each other, they will take less storage room.

Don't be misled by initial costs. If you are serious about freezing, buy containers that can be reused. They will be less expensive in the long run.

Here is a list of acceptable containers for freezing fruits and vegetables, along with tips for using them, to consider carefully before you buy:

1. Rigid containers of aluminum, glass, plastic, tin, or heavily waxed cardboard. Seal with lids that either press or screw on. If necessary, use freezer tape to make them airtight.
2. Glass canning jars, except for foods packed in liquid. Seal according to manufacturers' instructions.
3. Polyethylene bags, with or without accompanying cardboard carton for better stacking in freezer. Fill bag with food, press to remove as much air as possible, and seal by twisting and folding back top of bag. Tie with string, rubber band, or paper-covered wire strip.
4. Aluminum foil or freezer wrap. Seal with freezer tape.

Freezing Vegetables Generally all fresh vegetables, except those that are eaten raw, can be frozen. Those that don't freeze well include radishes, cucumbers, lettuce, scallions, and uncooked tomatoes. Some varieties freeze better than others, so if you plan to freeze check your State Extension Service for the names of the best varieties and buy seeds or plants accordingly.

Pick vegetables just before freezing time and use only those that are at their peak of ripeness. The chart that follows will give you an approximate idea of the amount of fresh produce necessary to yield one pint of a frozen vegetable.

SEVEN EASY STEPS TO FREEZING FRESH VEGETABLES

1. Sort vegetables according to size unless they are to be cut into pieces. Don't use anything that is not ripe or is damaged in any way.

2. Wash vegetables thoroughly in cool water. Lift them gently out of the water to let sand and grit settle to the bottom. Vegetables that are protected by pods, such as peas and lima beans, don't need washing.

3. Prepare vegetables as directed in the chart following.

4. One of the most important steps in freezing vegetables is blanching. Blanching means simply heating for a specific length of time in boiling water in order to stop the enzyme action; if not controlled, this will cause unpalatable changes in color, texture, and flavor. Blanching can be done in a special blanching pot, which comes with a basket and cover. If you don't have a blancher, use a wire basket and a large pot or kettle with a cover.

   Bring to a boil one gallon of water for every pound of vegetables.

   Place about a pound of prepared vegetables in the blanching or wire basket. If you don't have a kitchen scale, use the amount recommended in the chart on page 198.

   Lower the basket into boiling water. If vegetables float to the surface, use a fine wire mesh cover to keep them submerged.

   Cover the blancher or kettle and *immediately* start to count time. Cook over high heat so water stays at a boil. Use the chart for exact blanching time. If you live 5,000 feet or more above sea level, add one minute to the blanching time on the chart.

5. After blanching, quickly cool vegetables so they stop cooking. To cool, plunge them into ice water or very cold running water. Chill them for about as long as they were blanched or until they feel cool to the touch. Drain thoroughly.

6. Pack into freezer containers, allowing the headspace recommended in the chart. Vegetables that pack loosely, like broccoli spears or asparagus, require no headspace. (Headspace is the area between the top of the vegetable and the top of the container. It is necessary for the expansion that takes place during freezing.)

7. Seal and label containers with the name of the vegetable and the date packed. Freeze immediately in small batches at zero degrees or below. It's best to place containers to be frozen against freezer coils or plates. Leave space for air to circulate between packages.

Storage Time Keep vegetables frozen at zero degrees or below until ready to use. Most vegetables, if frozen properly, will maintain high quality for 8 to 12 months. Vegetables frozen for a longer period of time are not harmful to eat, but may be less desirable in flavor, texture, and appearance.

Label all frozen foods as to the contents and date of packaging, then keep a chart beside the freezer that indicates when foods are put in and removed. A complete turnover of food within one year is recommended and probably necessary to maintain high quality.

FREEZING CHART FOR VEGETABLES

Vegetable and Approximate Yield	Preparation and Freezing Directions	Amount to blanch per one gallon water	Blanching Time	
Asparagus 1–1½ lbs. fresh = 1 pint frozen	Sort spears according to thickness. Remove scales and wash thoroughly. Break off each spear as far down as it will snap easily. Leave spears whole or cut in 2-inch lengths. Blanch, cool quickly. Drain and pack as directed on page 197. No headspace necessary. Seal, label and freeze.	1 pound or 30–40 medium spears	Small spears Medium spears Large spears	2 minutes 3 minutes 4 minutes
Beans, Green or Snap ⅔–1 lb. fresh = 1 pint frozen	Wash and remove ends. Leave whole, cut in 1- or 2-inch pieces, or cut French style. Blanch, cool quickly. Drain and pack as directed on page 197. Leave ½-inch headspace. Seal, label, and freeze.	1 pound or 1 quart		3 minutes
Beans, Lima (*in pods*) 2–2½ lbs. = 1 pint frozen	Shell or leave in pods and shell after blanching. Sort beans according to size. Blanch, cool quickly. Drain and pack as directed on page 197. Leave ½-inch headspace. Seal, label and freeze.	1 pound or 3 cups	Small beans Medium beans Large beans	2 minutes 3 minutes 4 minutes
Beets (*without tops*) 1¼–1½ lbs. = 1 pint frozen (See **Greens** for beet tops)	Wash and sort beets according to size. Remove tops, leaving ½-inch of stem and all of root. Cook in boiling water to cover until tender. Cook small beets 25 to 30 minutes, medium beets 45 to 50 minutes. Cool quickly in cold water and drain. Peel and leave whole, slice, or cut in cubes. Pack as directed on page 197. Leave ½-inch headspace. Seal, label, and freeze.	Blanching not necessary		
Broccoli 1 lb. fresh = 1 pint frozen	Wash thoroughly, peel stalks, and trim off bottom of stem and large leaves. If necessary to remove insects, soak for ½ hour in a solution of 4 teaspoons salt and 1 gallon cold water. Split stalks lengthwise in uniform pieces so flowerets are not more than 1½ inches across. Blanch, cool quickly. Drain and pack as directed on page 197, alternating stems and flowerets. No headspace necessary. Seal, label, and freeze. Broccoli can be steamed instead of blanched before freezing. Steam 1 pound for 5 minutes. Cool quickly and pack as above.	1 pound or about 12 stalks		3 minutes

Brussels Sprouts 1 lb. fresh = 1 pint frozen	Remove any damaged leaves and wash thoroughly. If necessary to remove insects, soak for ½ hour in a solution of 4 teaspoons salt and 1 gallon cold water. Sort heads according to size. Blanch, cool quickly. Drain and pack as directed on page 197. No headspace necessary. Seal, label, and freeze.	1 pound or 3 cups	Small heads Medium heads Large heads	3 minutes 4 minutes 5 minutes
Carrots (*without tops*) 1¼–1½ lbs. = 1 pint frozen	Remove tops, wash and scrape or pare thinly. (Immature or small carrots need not be pared or scraped.) Leave small carrots whole and cut others into cubes, ¼-inch slices, or strips. Blanch, cool quickly. Drain and pack as directed on page 197. Leave ½-inch headspace. Seal, label, and freeze.	1 pound or 3 cups or 6–8 small carrots	Whole Cut	5 minutes 2 minutes
Cauliflower 1⅓ lb. fresh = 1 pint frozen	Cut or separate into clusters about 1 inch across. Wash thoroughly. If necessary to remove insects, soak for ½ hour in a solution of 4 teaspoons salt and 1 gallon cold water. Blanch, adding 4 teaspoons salt per gallon of water. Cool quickly. Drain and pack as directed on page 197. No headspace necessary. Seal, label, and freeze.	1 pound or 3 cups		3 minutes
Corn (*in husks*) 2–2½ lbs. = 1 pint frozen	ON THE COB: Don't use overmature corn. Husk, remove silk, and wash corn. Sort according to size. Blanch and cool quickly as directed on page 197. Drain. Pack into containers or wrap in moisture- and vaporproof wrap. Seal, label, and freeze. Corn on the cob takes up a lot of space, so don't freeze corn this way unless your freezer is large.	1 pound or 2–4 ears, depending on size	Small ears Medium ears Large ears	7 minutes 9 minutes 11 minutes
	CREAM STYLE: Husk, remove silk, and wash corn. Blanch and cool quickly as directed on page 197. Drain. Cut corn from cob at about center of kernel. Scrape cob to remove juice and heart of kernel and add to cut kernels. Pack, leaving ½-inch headspace. Seal, label, and freeze.	1 pound or 2–4 ears, depending on size		4 minutes
	WHOLE KERNEL: Husk, remove silk, and wash corn. Blanch and cool quickly, as directed on page 197. Drain. Cut kernels from cob about ⅔ depth of kernel. Pack, leaving ½-inch headspace. Seal, label, and freeze.	1 pound or 2–4 ears, depending on size		4 minutes

FREEZING CHART FOR VEGETABLES

Vegetable and Approximate Yield	Preparation and Freezing Directions	Amount to blanch per one gallon water	Blanching Time
Greens (*Swiss chard, beet tops, spinach, turnip tops*) 1–1½ lbs. fresh = 1 pint frozen	Wash thoroughly and remove tough stems and imperfect leaves. Blanch; cool quickly. Drain and pack as directed on page 197. Leave ½-inch headspace. Seal, label, and freeze.	1 pound or 1 quart	2 minutes
Parsnips 1¼–1½ lbs. fresh = 1 pint frozen	Remove tops, wash and scrape or pare thinly. Cut in slices or ½-inch cubes. Blanch, cool quickly. Drain and pack as directed on page 197. Leave ½-inch headspace. Seal, label, and freeze.	1 pound or 3 cups	2 minutes
Peas (*in pods*) 2–2½ lbs. = 1 pint frozen	Snap open pods and remove peas. Discard any starchy peas. Blanch, cool quickly. Drain and pack as directed on page 197. Leave ½-inch headspace. Seal, label, and freeze.	1 pound or 2½ cups	1½ minutes
Spinach	See **Greens**		
Squash, Summer 1–1¼ lb. fresh = 1 pint frozen	Wash and cut off thin slice at stem and blossom ends. Do not pare. Cut in slices ½-inch thick. Blanch, cool quickly. Drain and pack as directed on page 197. Leave ½-inch headspace. Seal, label, and freeze.	1 pound or 3 cups	3 minutes
Squash, Winter 1½ lb. fresh = 1 pint frozen	Wash and cut into uniform pieces. Pare; remove seeds and stringy portion. Cook, covered, in 1-inch boiling water until tender (about 15 to 20 minutes, depending on size of pieces). Drain. Or steam squash for 20 minutes. Mash or press pulp through a strainer or food mill. Cool in pan set in ice water. Pack into containers, leaving ½-inch headspace. Seal, label, and freeze.	Blanching not necessary	
Swiss Chard	See **Greens**		
Turnips (*without tops*) 1¼ to 1½ lbs. = 1 pint frozen (See **Greens** for turnip tops)	Remove tops, wash and pare thinly. Cut in ½-inch cubes. Blanch, cool quickly. Drain and pack as directed on page 197. Leave ½-inch headspace. Seal, label, and freeze.	1 pound or 3 cups	2 minutes

COOKING FROZEN VEGETABLES

The most popular way to get garden-fresh vegetables from the freezer to the dinner table is by boiling them without thawing. Exceptions are greens and corn on the cob, which should be partially thawed for more even cooking.

To cook one pint of frozen vegetables in water:

1. Bring one-half cup lightly salted water to a boil. (Use one cup water for lima beans, and water to cover for corn on the cob.)
2. Add frozen vegetables, cover, and quickly return to a boil.
3. Reduce heat and boil gently according to the table below. (Times should be increased slightly in high altitudes.)
4. Lift cover once or twice and separate pieces gently with a fork.
5. Once thawed, never attempt to refreeze vegetables.

BOILING TIMETABLE FOR FROZEN VEGETABLES

VEGETABLES		MINUTES (after water returns to boil)
Asparagus		5 to 10 minutes
Beans, Green or Snap	1-2 inch pieces	12 to 18 minutes
	French style	5 to 10 minutes
Beans, Lima		10 to 15 minutes
Broccoli		5 to 8 minutes
Brussels Sprouts		5 to 10 minutes
Carrots, cut		5 to 10 minutes
Cauliflower		5 to 8 minutes
Corn, cream style or whole kernel		3 to 5 minutes
Corn, on the cob		3 to 4 minutes
Parsnips		5 to 10 minutes
Peas		3 to 5 minutes
Spinach		4 to 6 minutes
Squash, summer		10 to 12 minutes
Turnips		8 to 12 minutes

Another popular method is to bake vegetables after they have thawed just enough so that pieces can be separated. Place them in a buttered ovenproof casserole and season as desired. Cover and bake in a preheated moderate oven (350 degrees) about 45 minutes, or until tender.

FREEZING FRESH FRUITS

Just think about the pure joy of having strawberry shortcake in the middle of winter and you'll be convinced that the freezer is one of the best places to store your fresh fruit. Special varieties of fruits have been developed to hold up better at freezing temperatures, so if you plan to freeze fruits, check with your State Extension Service for the names of these varieties and buy accordingly.

Pick ripe fruit just before freezing and pack it at once. See the chart on page 204 for the approximate amount of fresh fruit necessary to yield one pint of frozen.

FOUR EASY STEPS TO FREEZING FRESH FRUITS

1. Sort fruit and remove immature or overripe pieces. Discard any bruised or soft fruit.
2. Wash carefully in cold water. Work with small amounts at a time to prevent bruising.
3. Prepare according to chart on page 204. Generally fruit is prepared for freezing the same way it is prepared for serving. Do not use any galvanized cookware, iron utensils, or chipped enamelware in preparation.
4. There are four ways to pack fruit for the freezer: syrup pack, dry sugar pack, dry pack, and tray or loose pack. The method you use will depend on how you plan to use the fruit after it's frozen. Consult the chart on page 204 for the best way to freeze the fruit you are working with before making the final decision.

Syrup Pack Retains texture and flavor of fruit well; use this method for fruits that will be eaten without further preparation, like desserts or chilled fruit in syrup.

1. Prepare the proper sugar syrup (see following chart); chill. You will need about one-half to two-thirds cup of syrup for each pint of fruit. The chart on freezing fruits (see page 204) will tell you the type of syrup recommended for each fruit. Bring water and sugar to a full, rolling boil. Remove from heat and cool. Refrigerate until ready to use.
2. Wash and prepare fruit according to chart on page 204. Chill fruit for speedier freezing.
3. Use rigid, moistureproof-vaporproof containers for packing. Spoon fruit into containers and pack tightly.
4. Pour cold syrup over the fruit—this speeds freezing. Leave correct headspace for the kind of container you are using (see chart).
5. Place a small piece of crumpled plastic wrap or aluminum foil on top of fruit to keep it submerged in syrup.
6. Seal containers using manufacturer's directions. For a good seal, the sealing edges must be free of any food or moisture.
7. Label containers with name of fruit, date packed, and method used.
8. Freeze immediately in small batches at 0 degrees or below. Leave room for air to circulate between the containers being frozen.

SUGAR SYRUPS
(percentages)

SYRUP TYPE	SUGAR	WATER	YIELD
40%	3 cups	4 cups	5½ cups
50%	4¾ cups	4 cups	6½ cups

SYRUP AND DRY SUGAR PACK HEADSPACE

Wide Top Opening	Pint ½ inch Quart 1 inch
Narrow Top Opening	Pint ¾ inch Quart 1½ inches

Dry Sugar Pack Retains texture and flavor well. This is best to use for fruits that will be cooked, since there is little liquid.

1. Place about one quart of prepared fruit in a large bowl or shallow pan.
2. Sprinkle with the amount of sugar recommended in the chart on freezing fruits.
3. Mix gently with a large spoon or pancake turner until most of the sugar has dissolved and some juice has been drawn out of the fruit. Avoid crushing the fruit.
4. Fill containers with sugared fruit. Leave the correct headspace for the kind of container you are using (see headspace chart above).
5. Place a small piece of crumpled plastic wrap or aluminum foil on top of fruit to keep it submerged in juice.
6. Seal and label containers with name of fruit, date packed, and method used.
7. Freeze immediately in small batches at 0 degrees or below. Leave room for air to circulate between the containers.

Dry Unsweetened Pack May yield a lower-quality product than fruit packed in sugar or syrup; used primarily for fruits that will be cooked or will be served to the calorie-conscious or those on low-sugar diets.

1. Spoon fruit into container, leaving one-half-inch headspace between top of fruit and top of container.
2. Seal and label containers with name of fruit, date packed, and method of packing used. Freeze immediately at 0 degrees or below, leaving space for air circulation between containers.

Tray or Loose Pack This is another method for packing fruits without using sugar. It differs from the dry pack in that fruits remain loose instead of freezing in a block. This is a terrific advantage when you want to use only some of the fruit and put the rest back in the freezer for another time.

1. Spread fruit in a single layer on a shallow tray or pan and freeze at 0 degrees or below.
2. When fruit is frozen, pack into containers and seal. (Headspace is not necessary because the fruit is already frozen and expansion has taken place.)
3. Label containers with name of fruit, date packed, and method used. Return to freezer immediately, leaving room for air to circulate between containers.

Storage Time If fruits are properly frozen and stored at a temperature of 0 degrees or below, they will maintain high quality for 8 to 12 months. Eating fruits stored longer than the recommended time is not harmful, but the quality of the fruit may suffer. Fruits packed unsweetened, without syrup or sugar, will lose quality sooner.

USING FROZEN FRUITS

All frozen fruit must be thawed before using. Thaw only as much as you plan to use at one time. Fruit packed in dry sugar thaws faster than that packed in syrup

FREEZING CHART FOR FRUITS

FRUIT AND APPROXIMATE YIELD	PREPARATION	PACKING
Blackberries 1⅓–1½ pints fresh = 1 pint frozen	Sort berries and remove any immature or overripe fruit. Wash and drain thoroughly.	*Dry Sugar Pack* (see page 203) Add ¾ cup sugar to 1 quart berries. Mix carefully to avoid crushing. Pack into containers, leaving recommended headspace. Seal and label. Freeze as directed.
Blueberries 1⅓–1½ pints fresh = 1 pint frozen	Sort berries and remove any leaves and stems. Discard all immature or overripe fruit. Wash and drain thoroughly.	Not recommended.
Raspberries 1 pint fresh = 1 pint frozen	Sort berries and remove any immature or overripe fruit. Wash and drain thoroughly.	Add ¾ cup sugar to 1 quart berries. Mix carefully to avoid crushing. Pack into containers, leaving recommended headspace. Seal and label. Freeze as directed.
***Rhubarb** ⅔–1 lb. fresh = 1 pint frozen	Cut off and dispose of leaves; wash stalks; remove any bruised areas. Cut in 1- or 2-inch pieces, or in lengths to fit container. Rhubarb may be heated in boiling water for 1 minute and cooled in cold water promptly before packing. This brief cooking helps retain color and flavor.	Add 1 cup sugar to 4 or 5 cups rhubarb. Pack and freeze as directed.
Strawberries ⅔ quart fresh = 1 pint frozen	Sort berries and remove any immature or overripe fruit. Wash and drain thoroughly. Remove hulls. Berries may be left whole, halved, or cut in slices.	Add ¾ cup sugar to 1 quart berries. Mix well. Place in containers, leaving recommended headspace. Seal and label. Freeze as directed.

* Although a vegetable, rhubarb is treated as a fruit when freezing it.

and fruit packed without any sweetening takes the longest to thaw.

If you are going to serve the fruit raw, thaw it in the unopened container, for best color and flavor, and serve while there are still a few ice crystals in it. If the fruit is to be cooked, thaw just until the pieces can be separated. Be sure to allow for any sugar that was added before freezing when using in a recipe.

Here are three methods of thawing and the approximate times to allow for thawing a pound of syrup-packed fruit:

6 to 8 hours in the refrigerator
2 to 4 hours at room temperature
½ to 1 hour in bowl of cool water

Syrup Pack (see page 202)	*Dry Pack* (see page 203)	*Tray or Loose Pack* (see page 203)
Place in containers. Cover with cold 40% or 50% syrup. Leave recommended headspace. Pack, seal, label, and freeze as directed.	Pack, seal, label, and freeze as directed.	Pack, seal, label, and freeze as directed.
Steam one minute, cool quickly. Pack into containers. Cover with cold 40% syrup. Leave recommended headspace. Pack, seal, label, and freeze as directed.	Pack, seal, label, and freeze as directed.	Pack, seal, label, and freeze as directed.
Place in containers. Cover with cold 40% syrup. Leave recommended headspace. Pack, seal, label, and freeze as directed.	Pack, seal, label, and freeze as directed.	Pack, seal, label, and freeze as directed.
Place in containers. Cover with cold 40% or 50% syrup. Leave recommended headspace. Pack, seal, label, and freeze as directed.	Pack, seal, label, and freeze as directed.	Pack, seal, label, and freeze as directed.
Place in containers. Cover with cold 50% syrup. Leave recommended headspace. Pack, seal, label, and freeze as directed.	Not recommended.	Not recommended.

CANNING FRESH PRODUCE

There's a big difference between a row of commercially canned tomatoes in the supermarket and a row of home-canned tomatoes on the pantry shelf. Anyone can buy tomatoes, but only serious gardeners and cooks can open a jar they have "put up" themselves. There is something enormously satisfying about eating a dish of tomatoes that you have grown and canned months before. Just ask someone who has done it.

Initially, canning runs into something of an investment, but once you own the equipment it turns out to be one of the most economical ways to preserve home produce. The equipment you need will depend entirely on the food you are working with and the method you use to process it.

THE TWO METHODS OF CANNING

The boiling-water bath and steam pressure are the two canning methods. In both, foods are heated to a temperature at which all organisms that are apt to cause spoilage are destroyed. This heating, or processing, is done either in a boiling-water-bath canner at 212 degrees or in a steam-pressure canner where the temperature reaches 240 degrees.

The boiling-water-bath canner is used to process all fruits, as well as tomatoes and pickled vegetables. The steam-pressure canner *must* be used for all vegetables except tomatoes and pickled vegetables.

CHOOSING THE RIGHT EQUIPMENT

Water-bath Canners A water-bath canner is essentially a large covered kettle with a rack to keep jars from touching the bottom. A special water-bath canner is not absolutely necessary; any large pot will do providing it has a tight-fitting lid and is deep enough to let the water boil at least two to four inches above the lids of the jars. A sturdy wooden or wire rack, such as a cake rack, can be used to keep jars off the bottom of the pot.

If you have a steam-pressure canner that is deep enough, it may also be used as a water-bath canner. In this case, cover but do not fasten the lid and leave the petcock wide open so steam can escape as the water boils.

Steam-pressure Canners A steam-pressure canner is a heavy kettle with a cover that can be locked to make the canner airtight. It is fitted with a special gauge so the high temperature necessary for processing can be reached. All parts of the canner must be clean and in good working order before using, and the gauge should be checked for accuracy before each canning season. Follow the manufacturer's directions for using the equipment.

Jars and Closures Several kinds of jars are used for home canning. They differ primarily in the method of sealing, and the choice is entirely one of personal preference. However, when selecting canning jars, keep in mind the size of your family and the number of servings each jar will hold.

All equipment must be in perfect condition for canning. Do not use any jars that are cracked or chipped, or any closures that are dented, rusted, or damaged, for these are unsafe.

Before using, wash jars in hot soapy water and rinse thoroughly. Wash and rinse all lids and bands. Follow the manufacturer's directions for preparing and handling specific equipment.

GENERAL CANNING DIRECTIONS

Canning is not difficult, but certain rules must be scrupulously followed to insure a successful and safe product. Each step is important. Never take any shortcuts.

Select young, fresh vegetables and fruits that have been harvested as close to canning time as possible. Sort according to size and discard any damaged pieces. The section that follows will give you an approximate idea of the amount of fresh produce necessary to yield one quart canned.

Wash vegetables and fruits thoroughly before cutting or breaking the skin. Wash small amounts at a time under cool running water or in several changes of water.

Prepare as directed in the following section.

Two methods are used to pack produce for canning, the raw or cold pack and the hot pack. The essential difference between the methods is simply that for the raw or cold method vegetables and fruits are packed uncooked, and for the hot pack they are precooked. Except for some vegetables, like greens and beets, which must be packed hot, the method you decide to use is entirely one of personal choice. It has been shown, however, that hot packing allows a tighter pack and provides a more nutritive product. In both methods vegetables are finally cooked in the jar.

RAW PACKING is very easy. The prepared raw produce is packed in jars and covered with boiling water.

HOT PACKING takes a little longer but is not difficult. First the vegetables and fruits are preheated in boiling water. Then they are packed in jars and covered with either the liquid in which they were preheated (cooking liquid) or more boiling water. Boiling water should be used if the cooking liquid is dark, gritty, or strong flavored, or if the amount is insufficient.

About Packing Some foods must be packed loosely in jars to allow for expansion; others can be packed tightly. It all depends on what you are canning and the method you are using. Follow the specific directions for packing each vegetable and fruit and the amount of headspace to leave between the top of the food and the top of the jar.

Immediately after filling jars, remove air bubbles by running a rubber bottle scraper or a table knife between food and side of jar. If necessary, add more liquid to cover food, but remember to leave enough headspace.

Wipe jar rims with a clean damp cloth and seal according to specific directions for the type of jar and closure being used.

Process fruits, tomatoes, and pickled vegetables in a boiling-water-bath canner and ali other vegetables in a steam-pressure canner. Use the times indicated in the following section for exact processing times.

How to Process in a Boiling-Water Bath:

1. Arrange filled, sealed jars on rack in canner that contains either hot or boiling water. If raw or cold-pack method is used, the water should be hot. If hot pack is used, the water should be boiling.

If necessary, add boiling water to bring the level of water one to two inches above tops of jars. Do not pour directly on jars.

2. Cover canner. Bring to a rolling boil and start counting the processing time. Water should boil gently and steadily during processing. If water boils away, add boiling water to keep jars submerged.

3. Immediately after processing, remove jars from canner and place on rack or folded cloth to cool. Leave space for air to circulate between cooling jars.

How to Process in a Steam-Pressure Canner:

These instructions may be applied to all steam-pressure canners. For variations in the specific equipment being used, follow the manufacturer's directions.

1. Pour two to three inches boiling water into canner.
2. Arrange filled, sealed jars on rack in canner.
3. Fasten cover securely.
4. Place canner over heat. When steam has poured steadily from vent for ten minutes, close petcock or put on weighted gauge.
5. When pressure rises to ten pounds (240 degrees) start counting processing time. Maintain pressure by regulating heat under the canner; do not open the petcock to lower pressure.
6. At end of processing time, remove canner from heat and allow pressure to drop to zero. After that, wait five minutes before opening the petcock or removing weighted gauge. Unfasten and remove cover, tilting it so steam escapes on the side away from you.
7. Remove jars from canner and place on rack or folded cloth to cool. Leave space between jars for air to circulate.

Approximately 12 hours after processing, all jars must be tested to make sure that they are tightly sealed. To test a jar with a flat metal lid, press the center of the lid; if it is down or stays down when pressed, the jar is sealed. If the top has metal screw bands to hold the lids in place, remove the bands. Other jars may be tested by tilting sideways and examining for leaks. Any jar that is not sealed tightly must either be used immediately or emptied, repacked, and reprocessed as if it were fresh.

Wipe jars and label each one with name of food and date. Store in a cool, dry, dark place for up to one year.

Once in a great while canned food becomes spoiled during storage. You probably will never encounter this, but just in case, you should certainly examine all canned food before using it. Before opening a jar, look for a leak or a bulging lid—

the signs of possible spoilage. Pull the rubber sealer out from the jar with pliers, or puncture the self-sealing lid, and lift up lid. Now watch for spurting liquid, an off color, an off odor, or mold. *Never taste food with any of these signs; discard suspect food immediately.*

Since some spoilage cannot be detected, you should take the following precaution with all home-canned foods before eating them (tomatoes are an exception). Bring the food to a rolling boil, cover, and boil at least ten minutes. Boil spinach and corn 20 minutes. If the food looks spoiled, foams, or produces an off odor during cooking, destroy it immediately without tasting.

CANNING SPECIFIC CROPS

ASPARAGUS

Approximate Yield:

2½–4½ lbs. fresh=1 quart canned

Preparation:

Wash and break off each spear as far down as it will snap easily. Remove scales and wash again. Leave whole or cut in 1-inch pieces.

Raw or Cold Pack:

Pack raw asparagus in jars as tightly as possible without crushing. Leave ½-inch headspace. Add ½ teaspoon salt to pints, 1 teaspoon salt to quarts. Cover with boiling water, leaving ½-inch headspace. Adjust lids and process in steam-pressure canner at 10 pounds pressure (240 degrees).

Processing Time:

pints—25 minutes; quarts—30 minutes

Hot Pack:

Prepare asparagus as directed and boil 2 to 3 minutes. Pack hot asparagus loosely in jars, leaving ½-inch headspace. Add ½ teaspoon salt to pints; 1 teaspoon salt to quarts. Cover with boiling cooking liquid. If liquid is gritty, discard and use boiling water. Leave ½-inch headspace. Adjust lids and process in steam-pressure canner at 10 pounds pressure (240 degrees).

Processing Time:

pints—25 minutes; quarts—30 minutes

BEANS, GREEN or SNAP

Approximate Yield:

1½–2½ pounds fresh=1 quart canned

Preparation:

Wash and remove ends. Cut in 1-inch pieces.

Raw or Cold Pack:

Pack raw beans tightly in jars, leaving ½-inch headspace. Add ½ teaspoon salt to pints, 1 teaspoon salt to quarts. Cover beans with boiling water, leaving ½-inch headspace. Adjust lids and process in steam-pressure canner at 10 pounds pressure (240 degrees).

Processing Time:

pints—20 minutes; quarts—25 minutes

Hot Pack:

Cover prepared beans with boiling water and boil 5 minutes. Pack hot beans loosely in jars, leaving ½-inch headspace. Add ½ teaspoon salt to pints, 1 teaspoon salt to quarts. Cover with boiling cooking liquid. If there is not enough to cover beans, add boiling water. Leave ½-inch headspace. Adjust lids and process in steam-pressure canner at 10 pounds pressure (240 degrees).

Processing Time:

pints—20 minutes; quarts—25 minutes

BEANS, LIMA

Approximate Yield:

3–5 pounds in pods=1 quart shelled, canned

Preparation:

Wash and shell beans. Wash again.

Raw or Cold Pack:

Pack raw beans loosely in jars. Do not shake or press down in jars. Leave 1 inch headspace from top of pint jar, 1½ inches from top of quart jar. Add ½ teaspoon salt to pints, 1 teaspoon salt to quarts. Cover beans with boiling water, leaving ½-inch headspace. Adjust lids and process in steam-pressure canner at 10 pounds pressure (240 degrees).

Processing Time:

pints—40 minutes; quarts—50 minutes

Hot Pack:

Cover prepared beans with boiling water, bring to a boil, and remove from heat. Pack hot beans loosely in jars, leaving 1 inch headspace. Add ½ teaspoon salt to pints, 1 teaspoon salt to quarts. Cover beans with boiling cooking liquid or boiling water. Leave 1-inch headspace. Adjust lids and process in steam-pressure canner at 10 pounds pressure (240 degrees).

Processing Time:

pints—40 minutes; quarts—50 minutes

BEETS

Approximate Yield:

2–3½ lbs. (without tops)=1 quart canned

Preparation:

Wash and sort beets according to size. Remove tops, leaving 1 inch of stem and all of root. Cook in boiling water to cover until skins slip off easily (about 15 to 25 minutes, depending on size of beets). Remove skins and trim. Leave small beets whole and cut large ones in slices or quarters.

Raw or Cold Pack:

Not recommended.

Hot Pack:

Pack hot beets in jars, leaving ½-inch headspace. Add ½ teaspoon salt to pints, 1 teaspoon salt to quarts. Cover with boiling water, leaving ½-inch headspace. Adjust lids and process in steam-pressure canner at 10 pounds pressure (240 degrees).

Processing Time:

pints—30 minutes; quarts—35 minutes

BERRIES (blackberries and raspberries; strawberries not recommended)

Approximate Yield:

1½–3 lbs. fresh=1 quart canned

Preparation:

Wash fruit and drain.

Raw or Cold Pack:

Use for raspberries. Pack raspberries into jars and cover with boiling syrup, leaving ½-inch headspace. Adjust lids and process in steam-pressure canner at 10 pounds pressure (240 degrees).

Processing Time:

pints—10 minutes; quarts—15 minutes

Hot Pack:

Use for blackberries. Add ¼ cup sugar to each pint of berries, ½ cup to each quart. Let stand until juicy. Bring to a boil in covered pan; shake pan to keep the berries from sticking. Pack into hot jars leaving ½-inch headspace. Adjust lids and process in steam-pressure canner at 10 pounds pressure (240 degrees).

Processing Time:

pints—10 minutes; quarts—15 minutes

CARROTS

Approximate Yield:

2–3 pounds (without tops)=1 quart canned

Preparation:

Wash and scrape or pare thinly. If carrots are small and young, they need not be scraped or pared and they may be left whole. Otherwise, cut in slices crosswise, cut lengthwise, or dice.

Raw or Cold Pack:

Pack raw carrots tightly in jars, leaving ½-inch headspace. Add ½ teaspoon salt to pints, 1 teaspoon salt to quarts. Cover with boiling water, leaving ½-inch headspace. Adjust lids and process in steam-pressure canner at 10 pounds pressure (240 degrees).

Processing Time:

pints—25 minutes; quarts—30 minutes

Hot Pack:

Cover prepared carrots with boiling water, bring to a boil and remove from heat. Pack hot carrots in jars, leaving ½-inch headspace. Add ½ teaspoon salt to pints, 1 teaspoon salt to quarts. Cover with boiling cooking liquid. If there is not enough to cover carrots, add boiling water. Leave ½-inch headspace. Adjust lids and process in steam-pressure canner at 10 pounds pressure (240 degrees).

Processing Time:

pints—25 minutes; quarts—30 minutes

CORN, CREAM STYLE

Approximate Yield:

1½–3 lbs. (in husks)=1 pt. canned

Preparation:

Husk, remove silk, and wash corn. Cut corn from cob at about center of kernel. Scrape cob to remove juice and heart of kernel and add to cut kernels.

Raw or Cold Pack:

Do not use quart jars. Pack raw corn loosely in pint jars. Do not shake or press down in jars. Leave 1½-inch headspace. Add ½ teaspoon salt to each jar. Cover with boiling water, leaving ½-inch headspace. Adjust lids and process in steam-pressure canner at 10 pounds pressure (240 degrees).

Processing Time:

pints—95 minutes

Hot Pack:

Do not use quart jars. To each quart of prepared corn add 1 pint boiling water, bring to a boil, and remove from heat. Pack hot corn in pint jars, leaving 1-inch headspace. Add ½ teaspoon salt to each jar. Adjust lids and process in steam-pressure canner at 10 pounds pressure (240 degrees).

Processing Time:

pints—85 minutes

CORN, WHOLE KERNEL

Approximate Yield:

3–6 lbs. (in husks) = 1 quart canned

Preparation:

Husk, remove silk, and wash corn. Cut corn from cob at about ⅔ depth of kernel.

Raw or Cold Pack:

Pack raw corn loosely in jars. Do not shake or press down in jar. Leave 1-inch headspace. Add ½ teaspoon salt to pints, 1 teaspoon salt to quarts. Cover with boiling water, leaving ½-inch headspace. Adjust lids and process in steam-pressure canner at 10 pounds pressure (240 degrees).

Processing Time:

pints—55 minutes; quarts—85 minutes

Hot Pack:

To each quart of prepared corn, add 1 pint boiling water, bring to a boil, and remove from heat. Pack hot corn loosely in jars, leaving 1-inch headspace. Add ½ teaspoon salt to pints, 1 teaspoon salt to quarts. Cover with boiling cooking liquid, leaving 1-inch headspace. Adjust lids and process in steam-pressure canner at 10 pounds pressure (240 degrees).

Processing Time:

pints—55 minutes; quarts—85 minutes

GREENS (Beet Tops, Chard, Spinach, Turnip Tops)

Approximate Yield:

2–6 lbs. fresh = 1 quart canned

Preparation:

Wash thoroughly and remove tough stems and imperfect leaves. In a large pot heat until wilted, using just enough water to prevent sticking, or steam in a cheesecloth bag 10 minutes.

Raw or Cold Pack:

Not recommended.

Hot Pack:

Before packing, cut through greens several times with a sharp knife to loosen them. Pack hot greens loosely in jars, leaving ½-inch headspace. Add ¼ teaspoon salt to pints, ½ teaspoon salt to quarts. Cover with boiling cooking liquid. If liquid is dark or gritty, discard and use boiling water. Leave ½-inch headspace. Adjust jar lids and process in steam-pressure canner at 10 pounds pressure (240 degrees).

Processing Time:

pints—70 minutes; quarts—90 minutes

PEAS

Approximate Yield:

3–6 lbs. (in pods) = 1 quart canned

Preparation:

Wash and shell peas. Wash again.

Raw or Cold Pack:

Pack raw peas loosely in jars. Do not shake or press down in jars. Leave 1-inch headspace. Add ½ teaspoon salt to pints, 1 teaspoon salt to quarts. Cover with boiling water, leaving 1-inch headspace. Adjust lids and process in steam-pressure canner at 10 pounds pressure (240 degrees).

Processing Time:

pints—40 minutes; quarts—40 minutes

Hot Pack:

Cover prepared peas with boiling water, bring to a boil, and remove from heat. Pack hot peas loosely in jars, leaving 1-inch headspace. Add ½ teaspoon salt to pints, 1 teaspoon salt to quarts. Cover with boiling cooking liquid or boiling water. Leave 1-inch headspace. Adjust lids and process in steam-pressure canner at 10 pounds pressure (240 degrees).

Processing Time:

pints—40 minutes; quarts—40 minutes

POTATOES, NEW WHITE

Approximate Yield:

2–3 lbs. fresh = 1 quart canned

Preparation:

Wash potatoes. Precook 10 minutes and remove skins.

Raw or Cold Pack:

Not recommended.

Hot Pack:

Do not use pint jars. Pack hot potatoes loosely in jars, leaving ½-inch headspace. Add 1 teaspoon salt to each quart. Cover with boiling cooking liquid or boiling water. Leave ½-inch headspace. Adjust lids and process in steam-pressure canner at 10 pounds pressure (240 degrees).

Processing Time:

quarts—40 minutes

RHUBARB

Approximate Yield:

2½–4½ lbs. fresh=1 quart canned

Preparation:

Wash and cut into ½-inch pieces.

Raw or Cold Pack:

Not recommended.

Hot Pack:

Add ¼ cup sugar to each pint of fruit, ½ cup to each quart. Let stand until juicy. Bring to a boil in a covered pan; shake pan to keep rhubarb from sticking. Adjust lids and process in steam-pressure canner at 10 pounds pressure (240 degrees).

Processing Time:

pints—5 minutes; quarts—5 minutes

SQUASH, WINTER, also PUMPKIN

Approximate Yield:

1½–3 lbs. fresh=1 quart canned

Preparation:

Wash and cut into uniform pieces. Pare; remove seeds and stringy portion. Cook, covered, in 1 inch boiling water until tender, about 20 to 25 minutes, depending on size of pieces. Drain and reserve liquid. Mash pulp or put through a strainer or food mill. If necessary, add some of the reserved cooking liquid to make pulp a little thinner than needed for pies.

Raw or Cold Pack:

Not recommended.

Hot Pack:

Pack hot pulp in jars, leaving ½-inch headspace. Adjust lids and process in steam-pressure canner at 10 pounds pressure (240 degrees).

Processing Time:

pints—65 minutes; quarts—80 minutes

TOMATOES

Approximate Yield:

2½–3 lbs. = 1 quart canned

Preparation:

Wash. Remove skins by scalding in boiling water for 1 minute, then quickly dipping in cold water. Cut out stem ends and slip off skins.

Raw or Cold Pack:

Pack raw tomatoes tightly in jars. Press down until spaces are filled with juice. Leave ½-inch headspace. Add ½ teaspoon salt to pints, 1 teaspoon salt to quarts. Adjust lids and process in boiling-water bath (212 degrees).

Processing Time:

pints—35 minutes; quarts—45 minutes

Hot Pack:

Cut prepared tomatoes into quarters. Bring to a boil, stirring constantly to prevent sticking. Remove from heat. Pack hot tomatoes in jars, leaving ½-inch headspace. Add ½ teaspoon salt to pints, 1 teaspoon salt to quarts. Adjust lids and process in boiling-water bath (212 degrees).

Processing Time:

pints—10 minutes; quarts—10 minutes

Appendix: Average Dates of Last Killing Frost in Spring and First Killing Frost in Fall *

The chart below gives a fairly good idea of when you can expect a killing frost. But remember, the shape of the land you live on can alter frost dates considerably even though you may live only a few miles away from one of the locations listed. If you live in a valley or near a large body of water, you may find your growing season much longer than a neighbor who lives on the top of a hill. If you are new to the area, verify the information provided here by checking with local farmers or asking residents who grow vegetables.

To use the chart simply turn to your state and find the city or town nearest you. Then follow the columns across for average date of last killing frost in the spring, average date of first killing frost in the fall, and the length (number of days) of the growing season.

Alabama	LAST FROST IN SPRING	FIRST FROST IN FALL	LENGTH OF GROWING SEASON (days)
Decatur	Mar. 26	Nov. 5	224
Gadsden	Apr. 1	Nov. 2	215
Tuscaloosa	Mar. 23	Nov. 9	231
Thomasville	Mar. 18	Nov. 10	237
Ozark	Mar. 7	Nov. 19	257
Citronelle	Mar. 8	Nov. 23	260
Arizona			
Flagstaff	June 3	Sept. 29	118
Holbrook	Apr. 30	Oct. 18	171
Parker	Mar. 2	Nov. 15	258
Phoenix	Feb. 5	Dec. 6	304
Tucson	Mar. 19	Nov. 19	245
Yuma	Jan. 12	Dec. 26	348
Arkansas			
Fort Smith	Mar. 21	Nov. 10	234
Hope	Mar. 21	Nov. 7	231
Jonesboro	Apr. 1	Nov. 4	217
Little Rock	Mar. 17	Nov. 13	241
Rogers	Apr. 10	Oct. 25	198
Warren	Mar. 27	Nov. 7	225
California			
Bakersfield	Feb. 27	Dec. 6	282
Brawley	Feb. 6	Dec. 10	307
Fresno	Feb. 10	Dec. 6	299
Napa	Jan. 22	Dec. 19	331
Paso Robles	Feb. 7	Dec. 7	303
Red Bluff	Apr. 9	Nov. 3	208

* Adapted from *Climate and Man,* Yearbook of Agriculture, United States Department of Agriculture, 1941.

Colorado	LAST FROST IN SPRING	FIRST FROST IN FALL	LENGTH OF GROWING SEASON (days)
Denver	Apr. 26	Oct. 14	171
Pueblo	Apr. 23	Oct. 14	174
Le Roy	May 5	Oct. 5	153
Las Animas	Apr. 29	Oct. 9	163
Grand Junction	Apr. 16	Oct. 24	193
Garnett	June 8	Sept. 12	96

Connecticut (See **New England)**

Delaware (See **Maryland)**

Florida			
Pensacola	Feb. 16	Dec. 13	300
De Funiak Springs	Mar. 2	Nov. 24	267
Apalachicola	Feb. 11	Dec. 13	305
Tallahassee	Feb. 25	Dec. 4	282
Gainesville	Feb. 22	Dec. 4	285
Jacksonville	Feb. 15	Dec. 11	299

Georgia			
Toccoa	Apr. 5	Nov. 3	212
Atlanta	Mar. 23	Nov. 9	231
Millen	Mar. 22	Nov. 10	233
Americus	Mar. 17	Nov. 15	243
Waycross	Mar. 9	Nov. 18	254
Thomasville	Mar. 6	Nov. 20	259

Idaho			
Boise	Apr. 23	Oct. 17	177
Mackay	June 1	Sept. 17	108
Moscow	May 6	Oct. 6	153
New Meadows	June 27	Aug. 27	61
Pocatello	Apr. 28	Oct. 6	161
Twin Falls	May 18	Sept. 26	131

Illinois			
Chicago	Apr. 13	Oct. 26	196
Mount Carroll	May 9	Oct. 4	148
Peoria	Apr. 15	Oct. 20	188
Springfield	Apr. 11	Oct. 22	194
Mount Vernon	Apr. 16	Oct. 21	188
Cairo	Mar. 29	Nov. 1	217

Indiana	LAST FROST IN SPRING	FIRST FROST IN FALL	LENGTH OF GROWING SEASON (days)
Bluffton	May 6	Oct. 8	155
Farmersburg	Apr. 19	Oct. 16	180
LaFayette	Apr. 27	Oct. 12	168
Mauzy	May 3	Oct. 7	157
Paoli	Apr. 25	Oct. 16	174
South Bend	May 7	Oct. 15	161
Iowa			
Alta	May 4	Oct. 8	157
Clarinda	Apr. 30	Oct. 10	163
Fayette	May 9	Oct. 1	145
Harlan	May 5	Oct. 5	153
Iowa Falls	May 5	Oct. 2	150
Oskaloosa	Apr. 29	Oct. 9	163
Kansas			
Columbus	Apr. 9	Oct. 24	198
Horton	Apr. 20	Oct. 16	179
Hutchinson	Apr. 20	Oct. 19	182
Colby	May 2	Oct. 9	160
Hays	Apr. 29	Oct. 14	168
Garden City	Apr. 25	Oct. 16	174
Kentucky			
Bowling Green	Apr. 14	Oct. 22	191
Earlington	Apr. 16	Oct. 20	187
Eubank	Apr. 22	Oct. 16	177
Irvington	Apr. 13	Oct. 21	191
Lexington	Apr. 16	Oct. 22	189
Maysville	Apr. 21	Oct. 19	181
Louisiana			
Amite	Mar. 18	Nov. 6	233
Calhoun	Mar. 23	Nov. 6	228
Cheneyville	Mar. 9	Nov. 11	247
Grand Cane	Mar. 12	Nov. 8	241
Houma	Mar. 1	Nov. 12	256
Jennings	Feb. 27	Nov. 20	266

Maine (See **New England**)

Maryland–Delaware	LAST FROST IN SPRING	FIRST FROST IN FALL	LENGTH OF GROWING SEASON (days)
Grantsville, Md.	May 20	Sept. 29	132
Frederick, Md.	Apr. 21	Oct. 17	179
Fallston, Md.	Apr. 20	Oct. 22	185
Cheltenham, Md.	Apr. 18	Oct. 20	185
Princess Anne, Md.	Apr. 21	Oct. 19	181
Milford, Del.	Apr. 19	Oct. 25	189

Massachusetts (See **New England)**

Michigan			
Grayling	May 27	Sept. 19	115
Harbor Beach	May 10	Oct. 10	153
Lansing	May 5	Oct. 10	158
Newberry	May 31	Sept. 22	114
South Haven	May 6	Oct. 15	162
Stambaugh	June 7	Sept. 9	94

Minnesota			
Duluth	May 10	Oct. 5	148
Minneapolis	Apr. 25	Oct. 13	171
Moorhead	May 8	Oct. 1	146
Morris	May 14	Sept. 27	136
New Ulm	May 10	Oct. 1	144
Winnibigoshish Dam	May 18	Sept. 24	129

Mississippi			
Batesville	Mar. 30	Oct. 28	212
Corinth	Mar. 28	Oct. 31	217
Kosciusko	Mar. 29	Nov. 2	218
Vicksburg	Mar. 8	Nov. 15	252
Meridian	Mar. 17	Nov. 11	239
Biloxi	Feb. 26	Nov. 29	276

Missouri			
Maryville	Apr. 22	Oct. 13	174
Steffenville	Apr. 23	Oct. 13	173
Columbia	Apr. 13	Oct. 19	189
Springfield	Apr. 8	Oct. 28	203
Koshkonong	Apr. 7	Oct. 29	205
Arcadia	Apr. 26	Oct. 13	117

Montana	LAST FROST IN SPRING	FIRST FROST IN FALL	LENGTH OF GROWING SEASON (days)
Billings	May 15	Sept. 25	133
Havre	May 11	Sept. 22	134
Helena	May 2	Oct. 2	158
Kalispell	May 5	Oct. 1	149
Miles City	Apr. 30	Oct. 5	158
Poplar	May 16	Sept. 18	125
Nebraska			
Geneva	Apr. 28	Oct. 9	164
Hartington	May 2	Oct. 7	158
Imperial	May 6	Oct. 4	151
North Loup	May 5	Oct. 4	152
Scotts Bluff	May 11	Sept. 26	138
Valentine	May 4	Oct. 3	152
Nevada			
Reno	May 8	Oct. 10	155
Winnemucca	May 11	Sept. 29	141
Elko	June 1	Sept. 12	103
McGill	May 26	Sept. 22	119
Tonopah	May 16	Oct. 12	149
Logandale	Mar. 19	Nov. 11	237
New England			
Orono, Me.	May 18	Sept. 25	130
Concord, N.H.	May 3	Oct. 3	153
Northfield, Vt.	May 21	Sept. 25	127
Boston, Mass.	Apr. 13	Oct. 29	199
Providence, R. I.	Apr. 17	Oct. 26	192
Hartford, Conn.	Apr. 19	Oct. 18	182
New Hampshire (See **New England**)			
New Jersey			
Atlantic City	Apr. 6	Nov. 7	215
Bridgeton	Apr. 20	Oct. 23	186
Dover	May 6	Oct. 8	155
Indian Mills	Apr. 27	Oct. 18	174
New Brunswick	Apr. 22	Oct. 20	181
Trenton	Apr. 13	Oct. 27	196

New Mexico	LAST FROST IN SPRING	FIRST FROST IN FALL	LENGTH OF GROWING SEASON (days)
Agricultural College	Apr. 6	Oct. 31	208
Albuquerque	Apr. 13	Oct. 28	198
Bloomfield	May 12	Oct. 6	147
Deming	Apr. 2	Nov. 1	213
Roswell	Apr. 7	Oct. 31	207
Taos	May 12	Oct. 5	146
New York			
Albany	Apr. 23	Oct. 14	174
Angelica	May 26	Sept. 24	121
Indian Lake	June 13	Sept. 5	84
Oneonta	May 14	Sept. 27	136
Oswego	Apr. 23	Oct. 24	184
Setauket	Apr. 11	Nov. 8	211
North Carolina			
Asheville	Apr. 11	Oct. 22	194
Charlotte	Mar. 18	Nov. 11	238
Edenton	Apr. 1	Nov. 4	217
Raleigh	Mar. 23	Nov. 9	231
Wilmington	Mar. 16	Nov. 17	246
Winston-Salem	Apr. 11	Oct. 25	197
North Dakota			
Bismarck	May 10	Sept. 27	140
Devils Lake	May 15	Sept. 23	131
Dickinson	May 20	Sept. 17	120
Fullerton	May 16	Sept. 24	131
Grand Forks	May 16	Sept. 25	132
Williston	May 15	Sept. 25	133
Ohio			
Cleveland	Apr. 16	Nov. 5	203
Findlay	May 4	Oct. 10	159
Columbus	Apr. 19	Oct. 23	187
Cambridge	May 7	Oct. 10	156
Cincinnati	Apr. 12	Oct. 25	196
Portsmouth	Apr. 20	Oct. 20	183

Oklahoma	LAST FROST IN SPRING	FIRST FROST IN FALL	LENGTH OF GROWING SEASON (days)
Kenton	Apr. 22	Oct. 19	180
Alva	Apr. 7	Nov. 1	208
Mangum	Mar. 26	Nov. 6	225
Oklahoma City	Mar. 28	Nov. 7	224
Muskogee	Mar. 26	Nov. 4	223
Durant	Mar. 22	Nov. 11	234

Oregon			
Portland	Mar. 6	Nov. 24	263
Corvallis	Apr. 15	Oct. 23	191
Roseburg	Mar. 30	Nov. 19	234
Baker	May 12	Oct. 3	144
Bend	June 8	Sept. 7	91
Klamath Falls	May 18	Sept. 26	131

Pennsylvania			
Coatesville	Apr. 25	Oct. 17	175
Towanda	May 7	Oct. 5	151
Emporium	May 20	Oct. 2	135
State College	May 2	Oct. 5	156
Uniontown	Apr. 28	Oct. 18	173
Franklin	May 16	Oct. 3	140

Rhode Island (See **New England)**

South Carolina			
Charleston	Feb. 23	Dec. 5	285
Cheraw	Mar. 29	Nov. 3	219
Columbia	Mar. 15	Nov. 18	248
Greenville	Mar. 27	Nov. 10	228
Greenwood	Mar. 24	Nov. 9	230
Kingstree	Mar. 15	Nov. 12	242

South Dakota			
Huron	May 4	Oct. 2	151
Watertown	May 16	Sept. 27	134
Sioux Falls	May 6	Oct. 3	150
Pierre	Apr. 30	Oct. 8	161
Camp Crook	May 20	Sept. 21	124
Rapid City	May 1	Oct. 4	156

Tennessee	LAST FROST IN SPRING	FIRST FROST IN FALL	LENGTH OF GROWING SEASON (days)
Jackson	Apr. 6	Oct. 25	202
Clarksville	Apr. 3	Oct. 26	206
Lewisburg	Apr. 10	Oct. 22	195
McMinnville	Apr. 12	Oct. 24	195
Decatur	Apr. 15	Oct. 23	191
Rogersville	Apr. 18	Oct. 21	186
Texas			
El Paso	Mar. 21	Nov. 14	238
Fort Stockton	Apr. 1	Nov. 13	226
Gainesville	Mar. 27	Nov. 8	226
Galveston	Jan. 21	Dec. 28	341
Paris	Mar. 20	Nov. 13	238
San Antonio	Feb. 24	Dec. 3	282
Utah			
Salt Lake City	Apr. 13	Oct. 22	192
Logan	May 7	Oct. 11	157
Manti	May 25	Sept. 26	124
St. George	Apr. 10	Oct. 23	196
Moab	Apr. 24	Oct. 10	169
Vernal	May 26	Sept. 21	118
Vermont (See **New England**)			
Virginia			
Callaville	Apr. 17	Oct. 22	188
Lynchburg	Apr. 4	Oct. 25	204
Norfolk	Mar. 19	Nov. 16	242
Richmond	Mar. 29	Nov. 2	218
Woodstock	Apr. 19	Oct. 18	182
Wytheville	Apr. 17	Oct. 16	182
Washington			
Sedro Woolley	Apr. 22	Oct. 22	184
Vancouver	Mar. 30	Nov. 11	226
Lakeside	Apr. 9	Oct. 25	199
Rosalia	May 13	Sept. 30	140
Sunnyside	May 2	Oct. 10	161
Walla Walla	Mar. 31	Nov. 5	219

West Virginia	LAST FROST IN SPRING	FIRST FROST IN FALL	LENGTH OF GROWING SEASON (days)
Glenville	May 2	Oct. 15	166
Hinton	Apr. 24	Oct. 23	182
Huntington	Apr. 17	Oct. 20	186
Morgantown	May 1	Oct. 13	165
Parkersburg	Apr. 18	Oct. 19	184
Romney	May 3	Oct. 7	157
Wisconsin			
Spooner	May 15	Sept. 26	134
Medford	May 18	Sept. 26	131
Waupaca	May 10	Oct. 3	146
La Crosse	Apr. 29	Oct. 9	163
Madison	Apr. 29	Oct. 17	171
Milwaukee	Apr. 22	Oct. 23	184
Wyoming			
Cheyenne	May 14	Oct. 2	141
Cody	May 17	Sept. 19	125
Green River	June 3	Sept. 11	100
Lander	May 18	Sept. 20	125
Lusk	May 26	Sept. 18	115
Sheridan	May 15	Sept. 22	130

Acid soil
When the reaction of the soil registers below pH 7.0 (neutral), the soil is acid. If soil becomes too acid (pH 4.5 or below) it will not support vegetable crops.

Aeration, of soil
A properly prepared and cultivated soil layer will permit the free flow of air through its pore spaces.

Aggregate, soil
A clump, block, or cluster of fine soil particles adhering to each other. A desirable crumbly soil contains numerous small aggregates.

Alkaline soil
A soil whose reaction registers above pH 7.0 (neutral) is said to be alkaline. Few soils contain so much alkali that crops won't grow.

Amendment, soil
Any material that tends to improve the condition of the soil and therefore makes it more productive. Lime, peat moss, and mulch are soil-improving materials.

Annual
A plant that cannot survive winters and must be sown from seed each spring.

Available water
In soil, the moisture that is free for absorption by plant roots.

Bacteria
Microscopic organisms, usually parasitic, that live in the soil and contribute to the breakdown of spent vegetable matter.

Balanced soil
Any soil containing in proper proportions all the elements necessary for healthy plant growth and development.

Banding
The placing of a narrow strip of fertilizer below or to one side of seeds or plants. This fertilizer is covered with a thin layer of soil to prevent direct contact with injury-susceptible seed.

Biological control
A system of insect control patterned after the systems in nature that maintain a balance of forces.

Broadcast
To scatter evenly over a large area, as in applying a balanced fertilizer to the soil's surface or sowing winter rye.

Chlorophyl
The chemical element in plants that is responsible for their green color and contributes to the process of photosynthesis.

Chlorosis
The condition in plants that results when too little chlorophyl is produced, usually due to a nutrient deficiency.

Clay
A fine mineral particle present in soils that, when in balance with other materials, contributes to good soil structure with favorable moisture-retaining qualities. Too much clay will cause compaction, slow down or stop drainage, and make soil difficult to work.

Cold frame
A frame or box placed directly on the soil or set into it, with a cover of a transparent material such as plastic film or glass. The cover can be hinged or completely removable. By slanting the cold frame toward the south, it captures direct warm rays of the sun and seeds can be started or tender crops hardened off before the growing season begins.

Companion crop
A specific crop that is planted next to or within a row of vegetables in order to discourage harmful insects. Nasturtiums, for example, planted with beans are said to discourage aphids.

Compost
Broken-down organic matter produced by an artificial process of layering biodegradable materials, soil, and manure or fertilizer. Often called synthetic manure, compost is high in nutrients and an invaluable material for improving soil structure.

Cover crop
A temporary crop planted toward the end of a growing season to cover the ground until spring. Usually one of the grasses—winter rye or oats or a legume such as vetch—a cover crop benefits the soil by preventing erosion and adds nutrients to the soil when turned under.

Crop rotation
A system of changing the location of crops in order to prevent injury due to soil-borne diseases and exhaustion of certain nutrients.

Crust
A thin layer of hard dry soil that forms over the topsoil during dry conditions. If soil tends to bake or harden, seeds should be covered with a mixture of peat moss with sand or soil so seedlings can sprout without resistance.

Cultivation
The process of mechanically stirring the surface soil in order to eliminate weeds as well as encourage soil aeration and water absorption.

Damping off
A fungus that attacks seedlings at soil level causing them to rot at the base, wilt, and die. This disease is more prevalent under moist airless conditions and in unsterilized soil.

Drainage
A soil that allows excess water to pass through it is said to possess good drainage. Since plant roots can drown as easily as they can wilt from drought, the garden soil should be open enough to provide for the free flow of water.

Drought
A long period of dryness during which some form of irrigation is necessary. Most gardeners can expect at least one dry spell during the growing season.

Ecology
The branch of biology that deals with the interrelationships between living things and their relation to the environment.

Erosion
The gradual wearing away of the earth's surface through the action of flowing water, wind, and other elements.

Foliar fertilizer
A material containing plant nutrients that is absorbed when applied to leaves.

Friable
Term for soil that breaks or crumbles easily when handled.

Fungi
Forms of plant life that do not possess chlorophyl and therefore are unable to produce their own food.

Furrow
A long, narrow, shallow trench in the soil created by a plow or other implement.

Green manure
Any crop, such as vetch, clover, or winter rye, that is grown primarily for the purpose of providing added nutrients when plowed under.

Hardening off
The process of gradually introducing plants that have been started indoors to outdoor conditions. Usually this is accomplished by holding plants in a cold frame for from seven to ten days before setting them out in the garden.

Hardpan
The hard-as-cement layer sometimes present just beneath the plowed layer of certain soils.

Hardy
Term usually applied to plants capable of withstanding moderate amounts of cold temperatures.

Humus
Well-decomposed, dark-colored organic material in soil. Humus is soil's most valuable element because it soaks up and stores nutrients and moisture.

Hunger sign
A discoloration or wilting of a plant leaf indicating a serious deficiency of a particular nutrient or nutrients.

Insecticide
Any synthetically prepared material designed to eliminate insects harmful to plants.

Intercropping
The practice of sowing a quick-maturing crop within or between the rows of a slower-growing crop in order to utilize all available space.

Irrigation
The artificial application of water to crop lands.

Leaching
The loss of nutrient materials caused by the draining of water through the soil.

Legumes
Plants of the family *Leguminosae* are especially valuable to the horticulturist because of their ability to "fix" free nitrogen, thereby making it available to plants. Examples of legumes: vetch, peas, beans, clover.

Lime, agricultural
A soil amendment used to counteract excess soil acidity. Common liming materials are ground limestone (calcium carbonate), hydrated lime (calcium hydroxide), and burned lime (calcium oxide).

Loam
The classification given to soil containing moderate amounts of silt, sand, and clay, specifically 7 to 27 percent clay, 28 to 50 percent silt, and proportionately less than 52 percent sand.

Manure
The refuse consisting of excreta of farm animals, often mixed with wood shavings or straw and rich in nitrogen. Most fresh manures must be "aged" before applying to garden areas. Processed animal manures are sold in bags at garden centers and are ready for use.

Microbe
A tiny living organism too small to be seen with the unaided eye.

Mulch
A man-made or naturally occurring layer of plant residue on the surface of the soil. Mulches are ideal for conserving moisture, preventing crusting, reducing soil erosion, controlling weeds, and improving soil structure. Sheet plastic, although hardly a plant residue, is often used to accomplish the same ends.

Nitrogen fixation
As a rule, the conversion of free nitrogen to nitrogen combined with other elements, either naturally or through the activity of soil organisms. These fixed forms eventually become available to plants.

Nutrients, soil
The 13 elements needed for healthy plants. Major nutrients: nitrogen, phosphorus, potassium; secondary nutrients: calcium, magnesium, sulfur; micro-nutrients: iron, boron, manganese, copper, zinc, molybdenum, and chlorine.

Organic matter, in soil
That portion of the soil consisting of plant and animal residues at various stages of decomposition, plus cells and tissues of a variety of soil organisms.

Peat
Organic matter only slightly decomposed that accumulates in the presence of excessive moisture.

Percolation
The downward movement of water through soil.

Perennial
A plant, usually nonwoody, that winters over, though dying to the ground, and sprouts each year without resowing.

pH
An arbitrary numerical scale reflecting the concentration of hydrogen ions in a particular soil. Ranging from 0 to 14 (with 7.0 representing a neutral soil), pH reaction indicates whether certain elements are available to plants because of the degree of soil acidity or alkalinity.

Photosynthesis
The process by which plants convert water and carbon dioxide into carbohydrates (sugars) in the presence of light. The presence of chlorophyl is necessary to convert light energy into chemical forms.

Pore spaces
The areas in soil that are not occupied by solid particles.

Predator
An animal or insect that depends on other animals or insects as its source of food.

Root crop
Any vegetable grown primarily for its edible root, as opposed to a leaf crop.

Sand
The coarsest of the three soil separates (others being silt and clay). Sand particles feel gritty to the touch and can be seen by the naked eye.

Seedbed
An area of carefully prepared soil free of sticks, stones, and hard plant residues where seeds can germinate under optimum conditions.

Side dressing
A light application of either an organic or chemical fertilizing material along both sides of a row of crops.

Silt
A mineral element of soil coarser than clay but not as coarse as sand. Moist silt will smear and feel slightly abrasive when rubbed between the fingers.

Soil sample
A portion of soil taken from various locations in the garden area for analysis purposes. A sample sent to a State Agricultural Experiment Station should indicate the previous crops grown, crops planned, and length of time since last liming or fertilizing.

Soil test
Any procedure used to determine the chemical characteristics of a soil, including pH reaction and available nutrients, for the purpose of designing a productive fertilizing program.

Space-yield ratio
The relationship between the amount of space a specific crop requires and the amount of crop produced. Corn, for example, yields much less per square foot of area than tomatoes.

Sphagnum
A genus of mosses that grow in moist humid regions. Yearly accumulations result in a fibrous, highly absorbent material.

Starting soil
Soil especially prepared for the purpose of providing seeds with a sterile, nutrient-rich medium for germination.

Sterilization, soil
The process of treating a soil either with extreme heat or with chemicals to eliminate undesirable disease-causing bacteria or spores.

Structure, soil
The particular arrangement of individual soil particles into clumps or clusters that characterizes a soil. Unstructured soil is either massive (made up of particles that stick together, tending toward a homogeneous mass) or single grain (composed of separate particles like sand). Neither is highly desirable for crops.

Stubble mulch
A mulch created by leaving on the soil plant residues of a previous crop in order to protect the soil during the winter months, until a new seedbed is ready to be prepared.

Subsoil
The layer of soil directly below the plowed or topsoil level. It contains little or no organic matter.

Succession planting
Early maturing crops, when harvested, may be replaced by new sowings of the same or other vegetables. In some climates even a third planting may have time to mature and be harvested.

Synthetic manure
Term often used to describe composted material that is high in nutrients and soil-building qualities.

Tender
Description for plants that cannot survive cold temperatures and must be planted after the ground has thoroughly warmed in the spring.

Thinning
The process of removing selected seedlings from an overcrowded row in order to provide adequate growing room for the remaining plants.

Topsoil
The thin film of dark surface material, ranging in depth from a few inches to several feet, which supports vegetable life. The topmost part usually contains humus and organic materials that have nutritive value.

Trace element
An obsolete term (replaced by micronutrients) for elements found only in small amounts in plants, but nevertheless sometimes essential to plant growth or animal health. See NUTRIENTS, SOIL.

Transplanting
The moving of any plant—from a seedling to a mature plant—to another location.

Index